A Century of
Solutions

A Century of Solutions

Jim H. Smith

GREENWICH PUBLISHING GROUP, INC.
LYME, CONNECTICUT

Produced and published by Greenwich Publishing Group, Inc.
Lyme, Connecticut

Design by Clare Cunningham Graphic Design
Essex, Connecticut

Separation and film assembly by
Scan Communications Group, Inc.

Library of Congress Card Number: 00-103126

ISBN: 0-944641-41-5

First Printing: May 2000

Photo Credits:

pp. 8-9:	courtesy of Robert H. Murphy
pp. 10-11, 16-19, 48 top:	©Bettmann/CORBIS
pp. 20, 36, 45 bottom, 61:	courtesy of John D. Murphy
p. 38	E. B. Luce photographer
p. 55	photography by Fred Hiss & Son
p. 61	photograph by Bruce Lindsay
p. 64	photograph by Drucker-Hilbert Co.
p. 65	photography by Ernest Lidell
pp. 66, 68-69	photograph by Arrow Commercial Photo Service, Inc.
p. 72 inset	photograph by Dowd, Wyllie & Olson Incorporated
p. 74, 107	photograph by Fred Ware
pp. 76, 86, 92, 101, 104	photograph by Meyers Studio Inc.
p. 79 top	official U.S. Navy photograph
p. 97	official Mystic Seaport photograph by Louis S. Martel
p. 104	courtesy of Victor P. Crafa
p. 105	courtesy of Irwin Corning
p. 106	courtesy of Harvey Roberts
pp. 126-127, 129, 133-138, 146-149,150-151:	photography by Jeff Hacket

All other images courtesy of The Wiremold Company.

Contents

"Almost Everything Happened to Us..."

But for a short, sweet statement — 44 words that spoke volumes — there was nothing particularly unusual about the printed program given to guests attending the luncheon on September 21, 1950, to celebrate D. Hayes Murphy's 50 years in the manufacturing field.

The tastefully produced programs were printed on heavy, textured ivory card stock with a deckle edge. Appropriately, D. H. Murphy's name and the words "Golden Anniversary" were embossed in gold.

What lifted the booklet from the realm of the commonplace was the brief salute centered on the inside left panel. Signed by "The Employees of The Wiremold Company," it read:

"It is with warm appreciation for the wise, just and kind leadership which D. H. has displayed throughout the years, that we congratulate him upon his 50th Anniversary and reaffirm our high regard and deep affection for him as our President and fellow worker."

Few captains of American industry could legitimately claim such loyalty. But loyalty, D. H. Murphy would have been quick to assert, was a cornerstone of his company's success. He was, above all, a man of principles, and he had earned the respect of his employees by treating them fairly and demanding the same consideration in return.

Unlike many company presidents, who sequestered themselves in their offices, he was hands-on, frequently visiting the production floors of the plant. In fact, there wasn't a single person working for The Wiremold Company who had not met him personally.

Earlier in the year, when the company's annual service awards had been presented, records showed that more than a third of the company's employees had been there at least 10 years. Nearly 75 employees had worked there for 20 years or more. The Wiremold Company was widely acknowledged to be a family company, an affirmation embracing not just the Murphy family, but also the many families represented in the workforce.

Saluting D. H. as a "fellow worker" was, thus, an acknowledgment that the old man was one of them. It was, perhaps, the highest honor The Wiremold Company's employees could have accorded him.

More than 300 people, practically everyone who worked for the company, attended the event. When, at last, they were seated and all eyes turned to the head table, what they saw was nothing less than living history.

The men seated alongside the founder included three representatives of the International Brotherhood of Electrical Workers (IBEW). They were Walter J. Kenefick, the union's international representative, and Edmond G. Goulet and Morris Johnson, respectively the vice president and the business manager of Local

1040. Also seated on the dais were Gebhart Schach, chairman of the Wiremold Foreman's Club, and four men who had come to Hartford 31 years earlier, when the company relocated from Pittsburgh — Secretary William D. Ball, Textiles Divisions Manager Charles E. Rutherford, Treasurer Louis S. Zahronsky and James M. Foley.

It was Walter Kenefick who began the platitudes. "This is a most memorable occasion," he said, "one that honors the founder of a very fine company.... You are indeed fortunate," he told the assembled employees, "to be associated with a man who has recognized the value of a labor union and who has found that recognition of what the labor movement stands for has given him the satisfaction of realizing a dream.... These workers have helped him attain that dream in its fullest. It has been a very happy association, and I hope we may again pay tribute to Mr. Murphy on many anniversaries to come."

Ball and Zahronsky followed Kenefick, the latter observing that material success was only one measure of Murphy's achievement. "There have been other, even more important evidences of your success," he said, "not the least of which has been the very fine reputation you have earned for yourself as a square shooter."

Finally, it was Murphy's time to speak.

Thunderous applause greeted him as he stepped to the microphone. He looked younger than his 73 years. Immaculately dressed in a gray, double-breasted suit he stood ramrod straight. As his clear eyes surveyed the sea of employees who had come to honor him, he saw many faces he had known for years.

The country was just five years removed from a war during which The Wiremold Company's support of the American military effort had led to innovations that would have unimaginable implications for the company's domestic business in the years to come. The *Hartford Times* would report that the typical American family was using four times as many electrical appliances as it had just three decades earlier. As Murphy would tell his employees later that afternoon as he reported on the company's profit-sharing plan, business in the last quarter had been good. But he also would caution against overconfidence.

When the applause subsided and he began to speak, his voice was clear, strong and energetic.

"No doubt 50 years seems like a very long time to most of you," he said. "And it is. But when I look back on those 50 years, they seem to have passed very quickly.

"It was a period of great progress in the world — progress in all the sciences except social science, the one that teaches us to get along with each other.

"There was peace — and war. Then peace, such as it was — and war again.

"There were good times and bad times — the biggest of all depressions and the biggest of all booms.

"In our struggling business, we had our little successes and our near-failures. Almost everything happened to us...."

chapter one

The Irish Boy

"Every business has a background,

a history. This business has

produced many a thrill and many

a heartache…. I am here to

tell you something about

how The Wiremold Company

started, where it's been…."

— D. Hayes Murphy,

Opening Session of The Wiremold School,

November 15, 1940

D. Hayes Murphy (standing, far left) was still at the dawn of his career when this family photograph was taken circa 1905. His father, Daniel E. Murphy (standing next to D. H.) had emigrated from Ireland and earned his own success in the insurance industry before helping D. H. get started in manufacturing. Daniel's wife (and mother of D. H.), Rosalie Maher Murphy, is seated in front of Daniel. To Daniel's left are D. H.'s sisters Tessie and Rosalie. Sisters Grace and Margaret are to the left of Mrs. Murphy.

John Falter's painting, *The Famished*, recalls the great Irish immigration begun by the devastating Potato Famine of the 1840s. Daniel Murphy came to America in 1859 in search of a fresh opportunity.

n 1859, when he was 16, the Irish boy looked back over some 12 generations of his family, and it seemed to him, in that long lineage, practically nothing had changed. When his ancestors first became tenants, little more than half a century had passed since Christopher Columbus made his historic voyage to America. Now, three centuries later, the family still didn't own the farm. Nor were they likely to own it.

If there was anything the boy was sure of, it was that he didn't want to grow old with cows. He wanted to make a better life for himself.

He had practically everything a smart lad needed to make it — a strong body, energy, ambition and a willingness to take risks. He was bright and adventurous. All he needed was a little money, and his family provided that.

They scraped together $100, wished him well and put him on a ship to America. He was determined to make a name for himself. And the name was Daniel E. Murphy. ■

The six weeks it took to cross the Atlantic was plenty of time for a young man with cash burning a hole in his pocket to make new friends.

The friends Daniel Murphy made were soldiers. They took him under their wing and taught him a few of the skills he would need to survive when he arrived in his new home. They introduced him to a game — all the rage in America — called poker.

That game must have seemed a metaphor for everything he expected to encounter in America — opportunity and tough competition. It was a game of will that could be won by a smart youth with some luck, some skill, the ability to bluff and the tenacity to endure.

Daniel Murphy left home with $100. He arrived in Boston with $1 and a new appreciation for the American way of doing things. It wouldn't be the first time in his life that he would lose money. But it would be the last time that he would lose money without having something to fall back on. ■

Destitute, Daniel Murphy slowly made his way from Boston to Connecticut, where he landed a menial job at a factory in Kensington. Though nearby New Britain would embrace

Kensington by the turn of the century and the city's sprawling industrial complex would become America's manufacturing capital (and a mecca for European immigrants), the cataclysm of the Civil War forced factory closings in the nearer term.

When the Kensington factory closed, Daniel moved to Hartford, where he learned the carpenter's trade and practiced it, with a journeyman's pride, for the next seven years.

In 1868, he moved to Bridgeport, where he invested his savings in a book and stationery store. It was a modest start, but he was determined to make it into something much grander. Over the next five years, he nurtured that small enterprise into a profitable real estate, life insurance and steamship agency, honing the business skills and building the fortune on which The Wiremold Company was founded.

By 1873, 14 years after arriving in America, the boy who had spent his first $99 learning the rules felt as if he had beaten the game. He had tucked away more than $10,000 — precisely what his next lesson in business would cost him. ■

The first rumblings of the economic earthquake that would soon roll through Connecticut began with the great Chicago fire of 1871. It and the Boston fire a year later cost America's growing insurance business a whopping $300 million. But the two fires were only the first cracks in the armor of the American economy that led to the Panic of 1873.

What broke the bank was overspeculation in railroad bonds. When Jay Cooke's vaunted Philadelphia banking house failed, it dragged the rest of the nation's economic structure down with it. By 1878, nearly every railroad in the country was in receivership, and more than 18,000 companies had gone down the sink. Daniel Murphy's was one of them.

But Murphy's native stubbornness had become hard muscle by now. Pulling himself up by the bootstraps, in 1874, he scraped together the remnants of his grubstake and, undaunted, headed west once more to find his fortune. ■

Daniel Murphy was 15 years and 3,600 miles from his homeland when he arrived in Chicago. What he found was a city unvanquished by its recent setback. Chicago was back on its feet and dusting off the ashes the way a game prizefighter — or a resilient Irish boy — shakes off an early knockdown. It was, in short, Daniel Murphy's kind of town.

The insurance industry was hard hit by the financial panic, but it would emerge from the recession as one of America's fastest growing industries. Here in Chicago, the rebounding economic epicenter of the Midwest, an experienced insurance agent such as Daniel Murphy had no difficulty finding work.

But he still felt a pull from the East. He had left behind Rosalie Gertrude Maher, of New Haven. He had met her during his years as a carpenter when he helped her father, a contractor, rebuild the New Haven railroad

Investors throng the San Francisco Stock Exchange in 1873, the year the bottom fell out of the U.S. economy. Daniel Murphy's fledgling business in Bridgeport, Connecticut, was one of the casualties of the financial panic, forcing Daniel to make a new career, this time in the Midwest.

Daniel Murphy joined the Northwestern Mutual Life Insurance Company in 1879. By 1903, the agency he headed in Milwaukee, Wisconsin, boasted more than $64 million of insurance in force, 10 percent of the company's entire book of business.

station after a fire. Once Daniel had regained financial stability, he returned for Rosalie.

They were married in 1875 and moved to Providence, Rhode Island, where their first son, Daniel Hayes, was born two years later. But the Midwest and the insurance business had made a strong impression on him.

In Rhode Island, he continued a correspondence he had begun in Chicago with the Northwestern Mutual Life Insurance Company. In the summer of 1879, the company's vice president offered him a job with the company in Milwaukee. Accompanied by his wife and his young son, Daniel Murphy pulled up stakes and headed west once more — this time, for good.

On June 5, 1882, Northwestern appointed him general agent for Michigan's northern peninsula. In 1887, he returned to Milwaukee and a larger territory. By 1903, his agency could boast more than $64 million of insurance in force. That figure represented more than 10 percent of the company's entire book of business and more than half of all the insurance in force in Wisconsin, where he competed with 34 other insurers.

When he died in 1906, Northwestern Mutual, in a tribute in one of its agent publications, proclaimed him "one of the half-dozen most successful agents in this country." ■

Daniel would not be able to pass on that legacy to his son D. Hayes. Try as Daniel might to interest him in the insurance business, the younger Murphy had other aspirations.

Known sometimes as Hayes or simply D. H., he had attended public schools in Milwaukee and had worked for his father during summers while attending the University of Wisconsin. But he had never found the insurance business interesting. "On the other hand," he would recall many years later, "I welcomed the opportunity to visit the Northwestern Malleable Iron Company," which was owned by the father of a classmate. "It was the comparison between the insurance company and the foundry that gave me my first idea that I would rather be in the manufacturing business. It seemed easier — at least until I got into it — and it certainly was more interesting."

Still, when D. H. graduated from college in "naughty-naught" (1900), his father created a job for him in the agency. "I had many talks with my father about what I was going to do," he would recall many years later. "I wanted to go into the manufacturing business. He was interested in having me follow in his footsteps."

Late in the fall, his father gave him an assignment.

"Go on up to Calumet," Daniel told his son, "and see what you can do. In the meantime, I'll be on the lookout for an opportunity in the manufacturing business for you."

D. H. took a company advance of $50 for traveling expenses and headed for the Michigan mining country to try his hand at selling insurance. He boarded a train to Calumet, arriving at two o'clock in the morning on November 14, 1900. When he stepped off the train, it was, he recalled, "a beautiful moonlit night, 40 degrees below zero."

Daniel's office had supplied him with a list of prospects. He got a good night's sleep and then promptly set about calling them.

The first man he called on said he was too old for insurance. The second man most assuredly was too old. He had died. The third man informed the youthful sales-man that nobody would be interested in talking with him on payday. He went back to his hotel and started making plans for the next day.

D. H. continued to call on prospects, but his luck didn't improve. Soon his funds began to dwindle and he decided he'd better find a job. Before he had a chance, however, he received a typically terse telegram from his father. It read: "Make gigantic effort to close all prospects and come home."

When he returned to Milwaukee, he learned that his father was "looking over" a manufacturing company for sale in Milwaukee. It was called the Richmondt Electric Wire Conduit Company, and the owner of the company was asking $12,000. The plan was for D. H. and the

owner of the Richmondt company to take a trip together. They would visit companies that bought Richmondt con-duit. D. H. would assess the status of the business and then report back to his father whether it was a good investment.

The first year of the new century was coming to an end, and The Wiremold Company was about to be born. D. Hayes Murphy had not survived long in the insurance business, but he would spend better than seven decades in his next enterprise.

The elegant Murphy home in Milwaukee reflects Daniel Murphy's success. The Irish boy who had come to the U.S. with only a dollar eventually excelled in the career of his choice and gave his son, D. H., the opportunity to do the same.

chapter two

Tenacity and Luck

"Our experience in this particular type of conduit dates back to the year 1900. The officers of our Company have been actively and continuously engaged in its production from that time until the present day."

— Advertisement for American Conduit

Manufacturing Company's Galvanite rigid conduit, 1913

In New Kensington, Pennsylvania, shortly after the turn of the century, workers in the machine shop of the American Conduit Manufacturing Company create the foundation on which The Wiremold Company would be built. At the time, the company produced several types of electrical conduit, including electro-galvanized steel pipe, enameled steel pipe and a popular loom product.

The engines turning the turbines of the largest electrical power plant in the world, right, cranked out 30,000 horsepower to drive 37 miles of the Manhattan Electric Railroad in 1902. Americans were eager to try the many "modern uses" of electricity, as observed by the Electric Club of New York City, a precursor of the National Electrical Manufacturers Association (NEMA), in this early twentieth-century advertisement, opposite.

As America left the nineteenth century behind, there was probably no better symbol of the nation's optimism about its technological future than electricity. By the turn of the century, building wiring systems had advanced so rapidly in response to growing consumer demand, that the industry was already evolving through several insulating technologies.

The wiring standard of the era, scarcely two decades removed from Thomas Edison's work at Menlo Park, was called "knob and tube." Two strands of electric wire were fastened three inches apart by means of porcelain knobs and then fed through the partitions and beams of buildings via a network of insulating tubes. In 1900, the state of the art in this technology was "loom," a woven, non-metallic tubing through which the conductor wires were pulled.

Though alternative insulators, especially a flexible metallic tube called armored cable, would soon displace loom's dominance as the insulation of choice in building construction, loom would remain an important form of electric insulation, especially in automotive applications.

But loom was not the product on which D. H. Murphy was to launch his career. Neither was armored cable. The Richmondt Electric Wire Conduit Company manufactured its own unique alternative, electrogalvanized metal conduit — rigid, zinc-coated tubes with a baked enamel lining on the inside.

It sounded complex, but "Richmondt's product was

1. Cluster Electrolier. 2. Setting Type Direct from the Improved Phonograph. 3. The Electric Combination. 4. A Storage Battery. 5. "Shining up" by Electricity. 6. Cooking on an Electric Stove. 7. Principle upon which an Electric Piano Works.

NEW YORK CITY.—THE MODERN USES OF ELECTRICITY—AN EVENING WITH THE ELECTRIC CLUB, NO 17 EAST TWENTY-SECOND STREET.

FROM SKETCHES BY A STAFF ARTIST.—SEE PAGE 235.

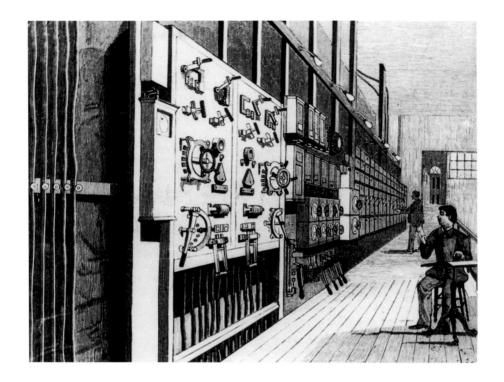

Less than 15 years after Thomas Edison perfected the light bulb, this 1893 woodcut shows the switchboard gallery of an Edison storage room in New York, unquestionably the epicenter of American electrification at that time.

simply water pipe enameled or painted on the inside and then used to protect electric wires used for household wiring," explains D. H. Murphy's son Robert.

It was a good product, but C. D. Richmondt didn't have much of a company. "The name was bigger than the company," Robert would wryly comment. It was an enterprise in trouble, which likely explained why its owner was prepared to sell it for a "mere" $12,000. And he clearly wanted to sell it to Daniel Murphy. Richmondt fully expected to sway young D. H. during the trip upon which they were about to embark.

Richmondt left his bookkeeper to watch over the business and climbed on a train with D. H. They would be gone for six weeks, and the neophyte industrialist would return a much wiser young man. ∎

Almost from the start, D. H. lost respect for Richmondt. The man's company was barely solvent, and it didn't take D. H. long to reach the conclusion that Richmondt was the primary problem.

Richmondt talked too much. And he was a dreamer. He envisioned all kinds of optimistic things for the company he was prepared to sell, but the $700 he withdrew to fund the excursion was nearly all the company had. Richmondt was certain he'd recoup the money, and much more, on the circuit of their customers. Murphy wasn't so sure.

Still, it was a valuable trip, if not for Richmondt's business, then for D. H. Murphy's business education. At Richmondt's expense, D. H. got a six-week introduction to the electric conduit industry. He also got a crash course in how *not* to do business. And it made an indelible impression on the young man.

The further they traveled, the more Richmondt confirmed D. H.'s conclusion that the thing he did best was spend money. It seemed they always ended up meeting with customers around midday, and inevitably Richmondt, who fancied himself a gourmand, would be the first to observe the noon hour and propose a lunch break. He seemed to know all the expensive restaurants. Over lunch, instead of talking business, he would hold forth on the virtues of fine dining. And The Richmondt

Electric Wire Conduit Company — whose rapidly dwindling liquid assets rested inside Richmondt's money belt — always picked up the tab. ■

In New York City, the duo met with "City" Brown, chief of the metropolitan department of water, gas and electricity. Landing him as a customer would mean a tremendous volume of business for the company.

"Brown was a typical city slicker," D. H. would remember. "We had no more than sat down when he began to pull out all kinds of junk," including samples of plumbing materials manufactured by other companies. "He kept on complaining [about these products]. Naturally, this disturbed me…"

Richmondt tried to explain that those products weren't his. But D. H. didn't think the message was getting through to Brown. The two men called on Brown several times without actually getting an order, and each time Richmondt would end up treating Brown and his staff to a lavish lunch. They ate well, but made no real progress with City Brown.

So it went, all across America. Like the man's style or not, the results seemed to speak for themselves. Over the course of the trip, Richmondt took orders for 23 carloads of pipe. It was an impressive volume of business, but there was a major downside. Richmondt didn't have the production capacity to deliver such volume.

By the time they returned to Milwaukee, Richmondt was out of cash and D. H. had decided two things. His father should buy Richmondt's company. And they should jettison Richmondt.

D. H knew he and Richmondt would never agree about how to run the business. Richmondt believed, for instance, that business was built on graft. D. H. discovered that Richmondt was taking a "rake-off" on practically all of the supplies the company bought and that one of the ways he got business was by crossing palms with silver.

Shortly after they had returned to Milwaukee, Richmondt received a letter from City Brown. "I received the sample which you sent me," Brown wrote, "but I was very much disappointed with the lining." When D. H. asked Richmondt about the letter, Richmondt shrugged and said, "You can't pay him enough."

"That was the first time I knew [Richmondt] was paying [Brown]," D. H. would later recall. "He had promised…$150 a month so that he wouldn't be complaining about all the junk that wasn't ours." The incident confirmed for D. H. that he and Richmondt could not work together.

Daniel Murphy, always a shrewd negotiator, paid $10,000 — $2,000 less than Richmondt's asking price — for the struggling company. As the year 1900 drew to a close, D. Hayes Murphy had embarked upon the business that would preoccupy him for the rest of his life. Papers filed with the State of Wisconsin named Daniel Murphy as The Richmondt Electric Wire Conduit Company's president and D. H. as secretary. Richmondt, who would soon be edged out of the business, was identified as vice president.

New York City, quickly wired for electricity after Edison demonstrated its utility as a light and power source, above, was an immensely important market for companies producing the materials required for electrification. D. H. Murphy made it clear that his company would not engage in unethical practices to gain that business but would do business on its own terms.

Wisconsin. 1900

Jessica Davis

Jessica Davis graduated in 1900 from Wisconsin University, where she met D. H. Murphy. The couple would marry in 1904.

Though D. H.'s father was reconciled to the fact that his son would never enter the insurance business, his mother had concerns of a different sort. She was worried about safety in manufacturing, and understandably so. Factories at the turn of the century were often dangerous places. D. H. took her concerns to heart. Worker safety would command his attention throughout his career.

The first thing D. H. did was to go on a second road trip and introduce the company's new management. "I didn't take $700 that time," he later reported, but he was particular to call on City Brown and explain that the new management would not be honoring the deal Richmondt had struck.

Brown greeted him cordially. And D. H., in turn, quickly spelled out the rules by which The Richmondt Electric Wire Conduit Company would be run now that he was in charge.

"We [don't] have 50 cents to pay anybody," he told Brown. "We are a small outfit, struggling to get started. We make an honest product and give good service to our customers. If we can't sell in New York on that basis, we'll have to pull out.…"

Those rules were, perhaps, the most immutable thing about the young company. D. H. would look back on that incident and remember it as the day when, less than two months into his career, he set the ethical standards by which his company would be operated. "I never heard from [Brown] again in my life," he said, "never paid him a nickel, and haven't…paid a nickel to anybody for graft. It just isn't done in *this* business." ■

Shortly after D. H. returned to Milwaukee from his second trip, one of Richmondt's customers came to query him. "I didn't know the answers," said D. H., "but I answered him. I saw his report afterward. It read something like this: 'Called on D. Hayes Murphy. He is very young. It is thought that his father, who has always scrupulously guarded his credit, will see the proposition through.'"

With not only his own reputation but also his father's on the line, it was make-or-break time for D. H. He had no choice but to give himself fully to the company. By November 1901, he had found a larger factory in Waukegan, Illinois, to which he could move the company in order to meet the unrealistic production Richmondt had promised to customers. Next, he got rid of Richmondt, who left with a substantial payout. D. H. always considered it a good investment.

Now D. H. was able, for better or worse, to take over the reins of the company. Nominally, he was secretary and treasurer. In fact, his job was to do whatever he had to do to keep Richmondt's poorly managed enterprise afloat.

He immediately changed the name of the company to something a bit more impressive — The American Interior Conduit Company. But, it was still mostly just a name. "There were no profits," he would remember, "just bills and bills and bills. At one time, I was spending my mornings in the office. In the afternoons, I would sleep. Then I'd get up and take on the job of night foreman."

It didn't take him long to realize that hard work

alone wouldn't solve his problems. Even with added production capacity in Waukegan, the company was getting killed by the competition.

So, in 1902, D. H. decided to stop fighting the competition and join it instead, merging American Interior Conduit with the largest conduit manufacturer in the business — the Safety Armorite Conduit Company of Pittsburgh. It is unclear what portion of ownership was transferred from the Murphy family to the Garland family, which owned most of Safety Armorite, but it is certain that D. H. no longer had a company of his own.

When the deal was consummated, John Garland, the head of Safety Armorite, took D. H. aside and said, "Here's the way I licked you. We had a big chart covering all territories and all discounts, with code words. When I would hear that you changed your price, I would telegraph Sears Conduit [with a five percent discount]."

The merger improved things a little. Safety Armorite operated The American Interior Conduit Company as its Waukegan branch. With Safety Armorite's production, D. H. could more comfortably deliver high volume — and he no longer had to do double duty as secretary/treasurer *and* graveyard shift foreman.

But he was still essentially a staff of one: keeping the books, learning to type letters by the "hunt and peck" system, answering the telephone, covering all the bases. "I had a very hard time of it," he remembered.

He would have a very hard time of it for two years. Little by little, though, things began to improve. In January of 1904, D. H. finally got some help in the office. With growing confidence in the business, he relaxed long enough to marry Jessica Esther Davis, who had been a classmate of his in college. Five months later, he moved the Waukegan operations, consolidating the company's manufacturing at Safety Armorite's West Pittsburgh facilities to be closer to the pipe supply. He was named general manager of the plant.

D. H. and Jessica bought a house in nearby New Castle, Pennsylvania, and Jessica moved into the company's offices with D. H. She took over the invoicing, relieving him of a huge volume of work. She was like a lucky charm, for also in 1904, the company would begin manufacturing Wireduct, the trade name for its loom product — an insulating tubing woven of cotton and twisted paper then impregnated with resin. It was an immediate success.

The next three years would be boom years for the company. ■

Early in 1905, a month after the Murphys' first wedding anniversary, their first child, Rosalie, was born.

D. H. was 28 years old, happily married and a new

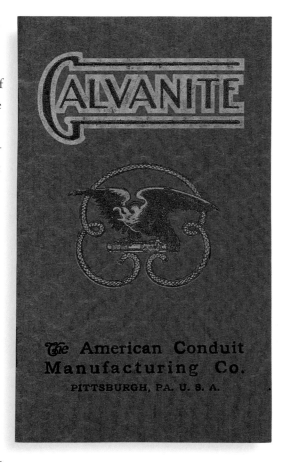

Galvanite rigid steel conduit was "pickled" to remove rust, coated with zinc and enameled on the inside to provide a perfect raceway for electric wiring.

Wireduct brand flexible conduit was introduced in 1904 with great success. Here women work the looms that produced the cotton tubing in American Conduit's loom facility in New Kensington, Pennsylvania.

father. His company was enjoying success, and he threw himself into his work with renewed enthusiasm. "I cut my eyeteeth in the Safety Armorite Company," he would say later.

But D. H.'s workload was complicated by internal struggles between competing factions of the Garland family, the owners of Safety Armorite. "I realized I was spending 80 percent of my time and energy combating intrigue in the organization," he said. "Everybody was trying to put something over on somebody else. At the end of the day, when I was all tired out, I'd have a chance to do a few strokes of business. I made up my

mind that if I were ever in a position to control policies, that sort of thing was something I would try to eliminate."

In addition to D. H.'s company, the Garlands also had purchased a separate company called the American Conduit Manufacturing Company, which they had run independently to prevent the public from perceiving them as monopolizing the market. "In those days, it was smart for a big manufacturer to have a small manufacturing firm and run it independently and as a competitor," D. H. said. "The customers were supposed to be suckers."

Early in 1906, the Garlands began to make a move that would consolidate a cluster of smaller companies, including the Murphys', under this umbrella. But by this time, D. H. had spent six years learning his business in some of the toughest classrooms imaginable. It hadn't been for nothing that he had wrested control of his company from Richmondt, worked double shifts as both bookkeeper and night foreman and survived the internal skirmishing at Safety Armorite. He refused to be squeezed out in the consolidation and even managed to secure some stock in the parent company.

"I made up my mind...that I was going to take care of myself," he said. ∎

A personal tragedy would complicate D. H.'s plan for his future.

Daniel Murphy died in 1906 after an extended illness. It had been nearly half a century since he had left his home in Ireland, arriving in America with only a dollar of the money his family gave him. He was 63 years old, and he had lived long enough to see his son become successful in the industry of his choice.

It had clearly been Daniel's intent to turn over his shares of the manufacturing business to D. H., and thus D. H. was not named in his father's will. However, his father inadvertently had never legally turned over his shares to D. H. Instead, shares of the company his father had intended for him were left to his mother, brother and sisters with the other assets of Daniel's estate. His mother got the family together to sort the matter out. The resulting redistribution of shares left D. H. with his own stake and the agreement of his family that he would manage the Murphy family's stake as a whole.

D. H. believed his greatest opportunity lay with the American Conduit Manufacturing Company. The Garland's handpicked president was ill, and was not likely to recover. As the Garland's looked for a successor, D. H. led the competitive scramble to fill the slot. D. H. was named secretary-treasurer of American Conduit and succeeded the company's president upon his death in 1909.

The first job he tackled as president was accounting. It was a discipline about which he had gotten a "boot camp" education over the past nine years. Still, he lacked the skills required for the task at hand.

"The accounts were in terrible shape," D. H. recalled, "a mess. The accounts receivable were a tremendous size in relation to sales. [The Garlands]

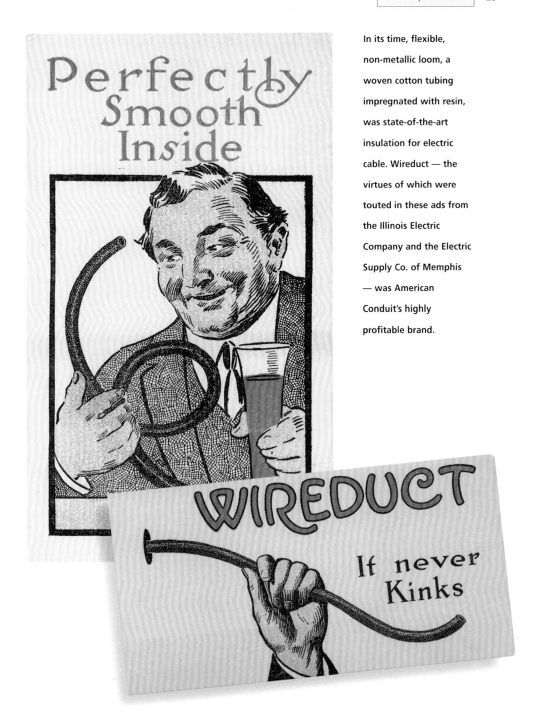

In its time, flexible, non-metallic loom, a woven cotton tubing impregnated with resin, was state-of-the-art insulation for electric cable. Wireduct — the virtues of which were touted in these ads from the Illinois Electric Company and the Electric Supply Co. of Memphis — was American Conduit's highly profitable brand.

had neglected to collect their accounts. I put a team of accountants on the job to clean house. After a while, I felt I was not getting what I wanted, and I got a second and finally a third firm of accountants. We eventually boiled down all the phony assets."

D. H. had no way of knowing how fortuitous this audit would turn out to be for him — or how well-timed. ■

D. H.'s young family was growing quickly. By now, Rosalie had a little sister, Jessica, born in 1906, and a baby brother, Daniel Hayes Murphy Jr., born in 1908.

But tragedy would visit the Murphys in 1909. In June, little Daniel fell ill with whooping cough and died. The bereft family decided it was time to leave New Castle. As soon as they were able, D. H. and Jessica bought a house in Pittsburgh, and D. H. set up his corporate offices downtown in the Keystone Building. Here they were closer to the new location of American Conduit's manufacturing operations in New Kensington, just outside of Pittsburgh.

D. H. had been working in his new role as president only three months when he got a call at home early one morning.

"The factory is on fire!" said the voice on the other end of the line.

"Which one?" D. H. asked.

"Both of them!"

The company had two production facilities in New Kensington. One was the rigid conduit plant, not much of a building, and as he jumped on a seven o'clock train,

D. H. had his fingers crossed that it was this building that was ablaze. The other facility was the company's loom plant, where the highly profitable Wireduct product was manufactured. Of the two products, it was clearly the more flammable.

Still, the Wireduct plant was 150 feet removed from the rigid conduit facility. D. H. couldn't imagine how they could both be on fire, and he was right. But when he got to New Kensington, his worst fears were confirmed. The loom plant was ablaze, and there was nothing he could do but watch it go up in smoke.

Fortunately, the building and its contents were insured. Production was halted only briefly while the plant was rebuilt. ■

His first decade in business had passed. The floundering company his father had purchased for $10,000 had grown into a major American enterprise.

But D. H. did not control his destiny. He had ceded that control to the stockholders of the Safety Armorite Conduit Company in 1902, and he knew that as long as he worked for the Garlands, he would never truly be a free man.

Then, in 1911, the Garlands got into serious financial troubles. And the problem was due entirely to their internal dissension, according to D. H. "That's why they went broke. They were fighting among themselves instead of fighting to protect the business."

But their misfortune was his opportunity to liberate himself. D. H. offered to buy the Garlands' share of the

The New Kensington facility is pictured here late in the winter of 1916. By this time, D. H. Murphy had freed the American Conduit Manufacturing Company from Safety Armorite. With a new mill, the company was already gearing up for the introduction of an innovative surface raceway that would revolutionize electrical wiring.

American Conduit Manufacturing Company at book value. Desperate, the Garlands accepted the offer. "Luck plays an awfully big part in these things," he would later explain. "I was lucky in two respects. In the first place, I was lucky to have boiled those assets down. In the second place, some stock I had received in the big merger the Garlands pulled was accepted by them at par in payment for the American Conduit Manufacturing Company. I was the only one who ever got a nickel out of that stock. The rest was absolutely worthless."

It was a transcendent moment, and it called for an expansive gesture — something symbolic. "Dad bought himself the biggest roll-top desk he could find and installed himself as president of the new company," Robert Murphy recalls.

D. H. Murphy was 34 years old, and he would never work for anyone other than himself again. From this foundation, he would build The Wiremold Company.

chapter three

Golden Pennies

*"When everybody works
and nobody shirks,
you can raise a firm from
the dead."*

— D. Hayes Murphy to employees in a
profit-sharing letter, 1917

age heralded by the emergence of more new gadgets — electric percolators, electric carpet sweepers, electric clothes washers. Before America would be able to fully enjoy all of these enhancements, however, there would need to be a critical change in the way American homes were wired.

American homemakers grasped the significance of the electric clothes iron practically from the moment it arrived. But, as illustrated in a series of classic advertisements for Wiremold raceway, there was a problem. As soon as homemakers got the new device home, many of them quickly discovered there was no place to plug it in. American homes had a woeful shortage of electric wall outlets. The ads depicted a frustrated homemaker unable to use her new appliance because the only place to plug it in was a light socket on the ceiling.

In fact, electric clothes irons had been available for fully 10 years when, in 1914, the American Conduit Manufacturing Company began tinkering with a new product specifically designed to address this problem. The new product was utterly unlike any of the others the company manufactured. Rigid conduit, loom, armored cable and non-metallic sheathed cable were all insulators designed to carry electric wiring through the walls of buildings. But, the new product would not run through walls. Instead, it represented a completely new idea in the management of wiring — a lightweight, molded metal conduit that could be easily installed *upon* walls and ceilings and *along* baseboards.

Charles W. Abbott, one of a number of bright young

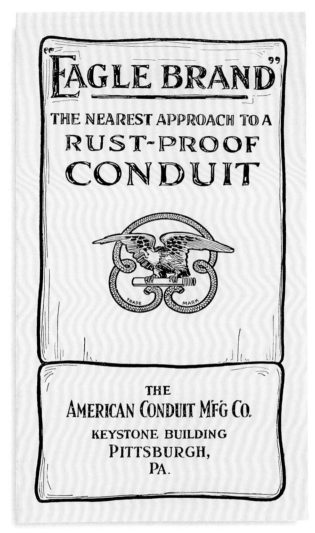

American Conduit introduced Eagle brand rigid conduit to capitalize on demand for this traditional product. But by 1914 American Conduit was already developing a unique alternative that would change everything.

professionals D. H. Murphy had hired, headed the team developing the new product. Abbott called it a "surface raceway." The concept was nothing less than brilliant. If homemakers could not reach the sparse electric outlets in their homes, the American Conduit Manufacturing

Company would make it easy for them to move the electrical outlets to more convenient locations.

It would take Abbott's team two years to perfect the design. By the time the new product debuted, the company was already imagining modifications for other applications. So the new product, with wiring capacity equal to one-half-inch conduit, got a series identification — the 500 series metal raceway, referring to the product's wiring capacity. (In the years to come, the company would introduce new raceways with different wiring capacities and design features under additional series numbers.)

But customers, who embraced it so enthusiastically that it remains a part of the company's product line to this day, would quickly come to know it by its product name — Wiremold. ∎

Wiremold raceway was introduced to the industry at the annual meeting of the National Electrical Contractors Association at New York's Hotel McAlpin in July, 1916. Some 250 contractors visited the Wiremold booth at the convention and their reaction to the new product was, Abbott reported, "instant approval. Many of them expressed their intention of immediately standardizing on it and many offered us initial orders. In fact, we could have sold 100,000 feet at this meeting if the National Electrical Code as then drawn had contained rules permitting its use."

D. H. was not concerned about the code. He had received assurances from individual members of the decision-making committee, as well as from officials at Underwriters Laboratories, Inc., (a major inspection firm), that the code changes needed to make sales of Wiremold raceway possible would be approved at the committee's biannual meeting the following spring.

Contractors liked the new product because it had been designed to correct many of the flaws inherent in earlier metal moldings that had been unsuccessful. "[None of] these materials proved popular," Abbott noted, "since, because of the characteristic variation in thickness of steel, uneven surfaces in buildings, etc. it was invariably difficult and frequently impossible to assemble their component parts out on the job without a great deal of difficulty. Wiremold was designed in the form of a flat pipe, or tube, and was made up of two pieces which might be permanently assembled at the factory in order to provide a simple, dependable means of coupling."

The Wiremold design, he noted, "affords a

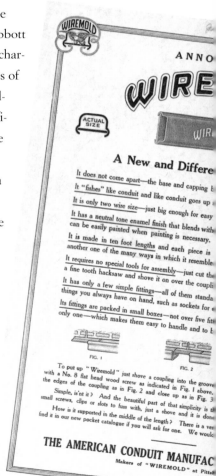

When Wiremold raceway was introduced to the industry at the annual convention of the National Electrical Contractors Association in New York City in June 1916, it met with immediate approval. Approximately 250 contractors visited the Wiremold booth in the Hotel McAlpin over six days.

Between 1909 and 1919, the Murphys made their home in industrial Pittsburgh, right. Above (left to right) are D. H., Rosalie, Jessica, Mrs. Murphy and Marjorie, with Robert (left) and John (right) seated on the floor.

simple, dependable electrical and mechanical connection between lengths and with fittings, without which any tube intended for surface wiring cannot succeed." Contractors needed two special tools to install the raceway product manufactured by National Metal Molding, a principal competitor. Wiremold raceway, on the other hand, could be easily installed with standard tools in every electrician's kit.

But its greatest competitive advantage was cost. National's product line had 158 different pieces. The Wiremold line was so well designed it needed only 15 fittings. Contractors didn't need calculators to appreciate the difference. It would cost them only $800 to stock Wiremold fittings, versus a whopping $7,000 for a comparable quantity of National's product.

Encouraged by the response he got in New York, Abbott launched an aggressive campaign in September to simultaneously sell the new product and win industry approval for its use. To the municipal electrical departments of New York, Chicago, Cleveland and other major cities where inspection was carried on under city ordinances, he sent extensive letters that explained, in elaborate detail, all of the advantages of Wiremold raceway.

Only one feature of the new product, in fact, was out of compliance with municipal codes. That was the "fishing" of wire into metal molding. But, he wrote, "since this feature has already been accepted and endorsed as an improvement by many of the large inspection departments — both Underwriters and Municipal — we trust that you too will be willing to waive it, as a point not anticipated by the code, pending formal approval."

At the same time that this letter was sent, Abbott launched a two-part promotional mailing to nearly 5,000 distributors and contractors. The letter included the endorsement of a contractor who had already tested Wiremold raceway and declared the product and its fittings "simple as a knife switch" and said it "costs less to buy and to install than any other surface raceway save wood molding."

Following this letter, they were sent actual samples of the "new two-wire metal molding — WIREMOLD by name — that is different from any metal molding you ever saw."

The response was nothing less than incredible. While a response in the range of 2 to 4 percent is considered good for today's direct-mail campaigns, Abbott's mailings generated responses from more than 36 percent of the contractors and distributors to whom it was mailed — in just the first three weeks! ■

Realizing that his company's success was directly related to the success of the electrical supply manufacturing industry, D. H. established himself as a prominent industry leader. In 1915, he became a founding member of the Association of Manufacturers of Electrical Supplies (AMES) and treasurer of its Rigid & Flexible Conduit Section.

He was, by now, frequently away from home on business. His young family, including his five children, would accompany him to Pittsburgh's Union Station when he left. Marjorie had been born in 1910, followed by John in 1911 and Robert in 1914.

"There was nothing more exciting than a ride to the downtown factory part of Pittsburgh at night so we could see all the bright fiery smoke coming out of the steel factory chimneys," the oldest daughter, Rosalie, would remember. "We thought it beautiful, and we called all the sparks golden pennies."

The sparks didn't look like coins to D. H., though. Increasingly, he yearned to leave Pittsburgh. If his children were captivated by the flames of industry, they were also suffering from recurring respiratory disorders. D. H. blamed these infections on Pittsburgh's industrial pollution.

He also missed sailing. In Wisconsin, there had been plenty of lakes to sail on. But Pittsburgh offered only dirty industrial rivers.

So, as he traveled around the country in search of new business and looking for ways to better organize the industry, he also began to look for a city where he could move his company. It would need to be a clean city, a good place to raise a family. It would need to provide skilled labor and a suitable factory building. And it would need to be near water.

Getting to the new city would not be easy, though. The company would need to be doing well, and D. H. would need to be able to transport its success quickly. He couldn't afford an extended lag in production. Wiremold raceway, he felt certain, would be his transportable product. ■

Ease of installation, cost effectiveness, simplicity of design and flexibility in application were billed as advantageous features of Wiremold raceway. Contractors heard the message. Here workers install the new product whose name would become synonymous with that of the company.

In advertisements, the Wiremold product name was, by now, dwarfing the company name: the American Conduit Manufacturing Company. By the time industry regulators met in March 1917 to revise the National Electrical Code, D. H. Murphy felt reasonably certain that — thanks to aggressive efforts to influence the opinions of industry leaders about Wiremold raceway — the company would have between $30,000 and $40,000 worth of orders

for the new product just as soon as it was approved.

In early 1917, he had sent distributors a special flyer titled "The Other Fellow," reminding them of the new product's successful introduction at the trade show in New York the previous spring. "You have been stung so often by wiring materials that looked good at first sight that you have dropped into the habit of letting The Other Fellow try out the new ones before taking them up yourself," it said.

"The Other Fellow likes Wiremold because it saves him money, because it saves him about thirty percent in labor, and because its fittings are so well standardized that he needs only about a dozen different styles for the average molding job."

The flyer went on to tell distributors that "you cannot afford to wait much longer to try out Wiremold yourself, because the time is soon coming when you are going to need metal molding and just now The Other Fellow has an edge on you. He knows he can make a fine job and a profitable one with Wiremold."

D. H. knew his market and, based on the response that Wiremold raceway had received at its New York introduction, he was counting on the flyer generating a flurry of inquiries just in time for the product's regulatory approval.

What he had not counted on was the Economy Fuse Company, a competitor that secured an injunction preventing the committee from approving Wiremold raceway. "[This injunction] left us minus approval, but with a large volume of business which we could not ship, and something over 1,500 contractors' inquiries which we felt sure we could cash in on with a relatively small amount of effort," he would recall a few years later.

If the Economy Fuse Company injunction came as a shock, no one was surprised when The National Metal Molding Company took advantage of the plight of Wiremold raceway. They lured back more than 20 contractors who had been prepared to change their allegiance as soon as Wiremold raceway became available.

National Metal's propaganda campaign was all the more damaging, as D. H. recalled, "since the trade papers published relatively little about the Economy Fuse Company's injunction proceedings and the great majority of contractors and wiremen took it for granted that the code meeting had been held and that we had failed to obtain approval for Wiremold."

All in all, it was a tremendous setback. It would be midsummer before Wiremold raceway finally got approved by Underwriters Laboratories and early 1918 before the revised code was actually distributed. Even then, American Conduit was obliged to mount an aggressive new marketing communications effort. Many building inspectors, confused about the approval, refused to permit installation of the new product until they could obtain a copy of the new code and review it for themselves.

"We were further handicapped," said D. H., "by the fact that The American Conduit Manufacturing Company operated almost exclusively through a few jobber agents and had only two representatives in the field, one in Philadelphia and one in Chicago, whereas the National Metal Molding Company was represented, either directly or through sales agents, by about 40 men spread from Boston to Los Angeles.

"As a result, we were forced to close our first real year of activity with Wiremold with a total billing of only about $60,000, nearly another $60,000 of business having been either cancelled because of this confusion over our

"The Other Fellow [the distributor who had already bought into Wiremold raceway] has an edge on you," this 1917 flyer warned electrical distributors. American Conduit was forced to repeat this massive marketing effort after the machinations of competitive companies stalled the product's approval.

The
Other Fellow

Housewives, enamored with the convenience of new electric appliances but frustrated by the shortage of electric outlets in their homes, would become a target audience for American Conduit's advertising. One of the earliest ads featured a series of photos illustrating the distressing limitations of inconvenient outlets on an otherwise trouble-saving device such as the electric iron. As illustrated in the promotional photo opposite, women in homes modified with Wiremold had no problem using their new electric appliances, including carpet sweepers.

new approval or never entered because [it was] conditional upon our securing our approval in March, as we had expected."

The company had introduced Wiremold raceway to tremendous acclaim. But thanks to legal setbacks, the initial momentum had been almost entirely lost. As D. H. put it, "We swung into 1918 faced with the necessity of regaining much ground and of practically making a new start." ∎

D. H. had reached the conclusion that the company would have to embark upon a national advertising effort. "We realized that in common with many other products, the Wiremold line was not one which would ever reach a maximum production through volume purchases in large quantity by a few jobbers, like rigid conduit," he said. "[It] was of the class where greatest possibilities lay in securing a large number of small orders from as many distributors as possible."

Trouble was, the company had practically no field organization. Instead, American Conduit Manufacturing Company was dependent on "our old jobbers agents in conduit and loom for exclusive control of Wiremold on the one hand and [faced] prejudice against us by as many other jobbers who felt certain we were giving our jobber agents an 'inside' on the other hand. We finally decided to venture into the national advertising field — this with a view to regaining the lost ground and gaining a great deal of new."

The "national advertising deal" was a brilliant

campaign launched with a direct-mail introduction sent to thousands of prospective customers coast to coast. "Even with our very limited sales organization," D. H. would later recall, "...we were able to merchandise our advertising campaign for about $50,000 worth of Wiremold business before the appearance of the first ad...."

Unfortunately, delays had put the company in a bad spot. With production slowed because of the war, American Conduit Manufacturing Company could only deliver about $6,000 worth of those first orders. Then, to make matters worse, the National Tube Company notified them that they could no longer provide the untreated pipe that American Conduit relied upon to manufacture its rigid conduit.

"It was special pipe, and our business had been built on it," D. H. recalled. "The National Tube Company just said they couldn't give us any more. [It was] a tragedy!"

The company had to find a new source. Considering that the war effort had reduced inventories of all raw materials, this was no small challenge. When they finally found pipe, it was of poor quality, so that the "pickling" process and the zinc coating process had to be doubled.

The company's efforts to address the setback, however, would be compounded by an even worse problem. Early in 1917, six of the company's accountants — practically the entire department — had enlisted in the U.S. Army. D. H. talked with the accountants. "They all assured me that the book inventory was correct and that

The Wiremold Selling Plan

"There is not one wired building in fifty in your town where sufficient outlets are to be found!" That was the bold declaration on which *The Wiremold Selling Plan* was predicated.

Packaged as a large format brochure, with a cover like an issue of the highly trusted *Saturday Evening Post*, the plan was sent to thousands of electrical contractors and distributors in 1918. Its aims were simple: to help them understand the enormous potential to make money retrofitting poorly wired buildings and to make it clear that the American Conduit Manufacturing Company was here to help them. The brochure both explained the company's plan to significantly increase Wiremold raceway's share of the market and showcased the high-impact advertising campaign the company was about to launch.

Not only did the lack of adequate wiring represent potential business for contractors, but fixing it was practically a moral imperative — especially when contractors had at their disposal an inexpensive, easy to install, utterly reliable product called Wiremold raceway.

And especially when, as the direct mailer made clear, *30 million* Americans would be exposed to the exciting new Wiremold advertising campaign in such popular periodicals as *Ladies Home Journal*, *Good Housekeeping*, and *Saturday Evening Post* "during the next few months."

To seal the deal, recipients of the mailer were offered "The Famous Wiremold Trial Order." For only $15.88, they could purchase 100 feet of Wiremold raceway and a package of each of the 15 fittings — "all that is necessary on the average job."

The Wiremold Selling Plan gave distributors a head start in selling Wiremold raceway, offering additional copies of the company's promotional literature and persuasive selling points.

they always made out requisitions for all material drawn. So I said, 'There's only one objective for all of us — that is to win the war. We'll go along with the book inventory if you fellows promise to be careful in every transaction.' In this way, he hoped to save his understaffed company the time and trouble of physically counting the inventory and thus maintain a brisk production rate.

Unfortunately, he soon discovered the accounting records were wrong. Some $40,000 worth of inventory he'd assumed was in the warehouse did not, in fact, exist. Overnight, the company's ratio of current assets to current liabilities plummeted to 1 to 1. "That's pretty hot water to be in!" D. H. would later recall.

It got hotter still. The Wiremold advertising campaign had generated some $30,000 worth of unfulfilled orders, "on every last cent of which customers were demanding immediate delivery." In addition, the company had almost 20,000 inquiries about the new product, and "more [were] coming in almost every mail and without the necessary clerical assistance to answer half of them or field assistance to so much as make a dent on cashing in on them," recalled D. H.

If American Conduit was to survive, he had to put Wiremold raceway on hold and focus on feeding the rigid conduit business. Years later he would write, "We were forced to stand by and see a truly remarkable amount of interest which could have been coined into business die out because the conditions of those first few months…[during] the war forced us to concentrate almost 100 percent of the production on rigid iron conduit."

There was another reason why D. H. was so resolved to keep the rigid conduit business healthy. He had identified an available property in Hartford, Connecticut, that would be ideal for his company's factory. The whole deal pivoted on his ability to sell the rigid conduit line so that the company could concentrate its resources on the expansion of its Wiremold raceway and loom products. D. H. even had a potential buyer — the General Electric Company.

"It was a desperate situation," D. H. would recall years later. "I placed all the facts before our bankers and was agreeably surprised to find that they appreciated our coming and said, "We'll work it out with you.""

With the bankers' support, D. H. did his best to keep the business solvent and General Electric interested. "We had notes on the open market," recalled D. H. "I always [remember] myself for nine months keeping open market paper up in the air with one hand [while] trying to get the General Electric Company to buy the plant."

In 1919, General Electric bought the rigid conduit business for $1 million. It gave D. H. the cash he needed to resolve the inventory discrepancy and move the company, and his young family, to Hartford.

The night before D. H. and his family left for Hartford, there was a big party at The Union Club, a popular Pittsburgh watering hole. It was the last day before Prohibition — the last night the club could sell its stock of liquor. D. H. came home from the party with a large stock of bottles that were then wrapped in various garments and stored in a wardrobe for the trip east. ∎

After selling the rigid conduit business to General Electric in 1919, D. H. moved his company to Woodbine Street in Hartford, Connecticut, as represented in the painting at left. Below is the Murphy family shortly after moving to Hartford: (left to right) Marjorie, Rosalie, D. Hayes, Mrs. Murphy, John, Jessica and Robert.

The company moved into the former Franklin Lamp Works building at 83 Woodbine Street in Hartford. It was a ponderous brick building, five stories high and sprawling more than a city block.

For the first time in three years, D. H. could breathe a little easier. The war was over. Prices were skyrocketing. American Conduit Manufacturing Company products were in demand. D. H. could not have guessed that he was about to make one of the worst mistakes of his career.

chapter four

The Wiremold Company

"How ya gonna keep 'em

down on the farm

after they've seen Paree?"

— popular American song, 1919

The former American
Conduit Manufacturing
Company had been The
American Wiremold
Company for four years
by the time the
company's sales force
gathered around
D. Hayes and Jessica
Murphy in front of
their West Hartford
home in 1924. D. H.
had restructured the
company's selling and
distribution arrange-
ments following the
move to Hartford to
better take advantage
of the construction
boom that would blaze
across the country in
the postwar years.

The S. S. *Belgic* passes the Statue of Liberty on January 15, 1919, bringing home World War I veterans full of the enthusiasm and optimism that would fuel the Roaring Twenties.

The nation's economy emerged from World War I like a newborn beast. For the briefest of moments, as though it could not be sure that this new, postwar world was real, it lay still, gathering its strength. Then it lurched to its feet, took a breath and almost instantly began to run.

And as it ran, it roared. Fueled by liberal credit and speculation that industrial inventories were going to grow, brisk sales in 1919 drove intense production, and the prices of everything skyrocketed. For the first time in years, the United States experienced a construction boom as Americans scrambled to make up for the building that had been neglected during the war.

American Conduit Manufacturing Company found itself swept along on the shoulders of this economic optimism. For a change, things were starting to go the company's way. Builders needed the raw materials for construction. And American Conduit would suddenly have a different problem — not a shortage of orders, but the inability to deliver the goods. ■

D. H. took advantage of the move to Hartford to dramatically alter American Conduit's marketing and distribution arrangements. Released from nearly all of the old obligations to jobber agents that the company had established in Pennsylvania, the company entered into affiliations with a number of prominent manufacturers' agents. For the first time, it was able to augment its effective representation in the South and the West with direct representation in many of the nation's largest metropolitan markets — the all-important New York, Philadelphia, Washington and Pittsburgh territories.

Staffed with seasoned and well-connected sales professionals, the manufacturers' agencies were ideally positioned to

WE TAKE PLEASURE IN ANNOUNCING THAT
MR. HARRY B. KIRKLAND
VICE-PRESIDENT
NOW HAS HIS PERMANENT HEADQUARTERS AT
71 WEST 23RD STREET
NEW YORK CITY
TELEPHONE. GRAMERCY 3723
MR. KIRKLAND IS IN CHARGE OF OUR NEW YORK DISTRICT SALES ORGANIZATION AND WILL DEVOTE HIS TIME EXCLUSIVELY TO THE SALE OF
WIREMOLD AND **WIREDUCT**
HARTFORD, CONN.
OCTOBER 1, 1931
THE AMERICAN WIREMOLD COMPANY

drive high volumes of business to the company. The company, however, was not in a position to deliver. "Despite our very strenuous efforts to avoid taking a large volume of Wiremold business during the period of our moving, and the difficult period in which we were organizing here in Hartford, we soon found ourselves once more oversold," D. H. would later recall.

Trying to meet this extraordinary demand for his products in a new factory and a new city, D. H. could not hire and train employees quickly enough. A disaster was practically inevitable. Producing loom at a feverish pace, the company inadvertently made a mistake. In the fall of 1919, it shipped a defective batch — a big batch. It was valued at over $80,000, roughly *$800,000* in 1999 dollars.

As the loom crossed America in railroad boxcars, the poorly prepared resin melted out of the fabric. By the time the shipment arrived at its destinations, the coils of loom had become clotted into useless wheels of tarry fiber. The entire shipment was sent back. And more than 150 customers demanded to know what American Conduit planned to do about the matter.

D. H. knew there was no way for the company to simply make up the loss to customers. For starters, there wasn't enough liquid capital on hand. Besides, if he simply paid back the money that was owed, he would miss an opportunity to reinforce the relationships he had worked so hard, and against such odds, to build. He was convinced that good customers, accustomed to good customer service, would be willing to overlook

the occasional mistake…if an honest effort was made to put things right.

D. H. made an exceptional offer. American Conduit Manufacturing Company would make good on the defective loom by doubling their orders. Just as quickly as possible, the company would deliver twice as much loom as the customers had paid for.

All but a handful of the customers accepted his largess, but it would be another four years before the company finished paying for this mistake. In the end it would be Wiremold raceway, increasingly popular thanks to innovative and aggressive marketing, that would begin to pick up the slack. ∎

Unprecedented postwar demand for loom and the challenges of a new factory would lead to disaster for the American Conduit Manufacturing Company. A batch of defective loom that shipped to 150 different customers forced D. H. Murphy to accept a staggering financial loss to keep those customers loyal. Harry B. Kirkland, who had helped ensure the company's independence nearly a decade earlier, moved to New York City in 1921 to help the company capture a greater portion of that rich market.

Any remaining question about the role D. H. expected Wiremold raceway to play in the company's future was addressed in 1920 when, still recovering from the loom disaster, he renamed his company for the second time in 20 years. Ensconced now in Hartford, it became The American Wiremold Company.

D. H. Murphy knew *exactly* what kind of product he had on his hands and what kind of potential market existed for that product. The fact that it wasn't achieving that potential — especially considering how hard the company had worked to market it — was more than frustrating.

Later that year, he assessed the lack of progress the company had made with Wiremold raceway in comparison to competitors. He wrote, "We are reliably advised that in 1917 the New York Territory of the National Metal Molding Company did an average net business on their metal molding of $25,000 per month and that during 1920 this net sales was almost doubled — the great bulk of this business being done in New York City where neither the company nor its metal molding...has been or now is popular with jobbers, contractors and wiremen as a class. In other words, as even a brief investigation will prove, National Metal Molding Company's sales...do not represent the possibilities open to a [popular] material like Wiremold in Greater New York and vicinity and that this is equally true of their sales in other territories.... Unfortunately, however, we have never made so much as a dent in the New York territory...."

Notwithstanding his concerns, the company's market-

ing efforts *were* paying off. He was obliged to concede that "despite the fact that circumstances prevented our plans to follow up the thousands of inquiries which came to us during our national advertising campaign...general trade interest in Wiremold has kept up remarkably."

THE AMERICAN WIREMOLD COMPANY

GENERAL OFFICES AND FACTORY

HARTFORD, CONN.

WIREMOLD WIREDUCT

BOSTON
NEW YORK
PHILADELPHIA
PITTSBURGH
ATLANTA
NEW ORLEANS
CHICAGO
DETROIT
DALLAS
LOS ANGELES
SAN FRANCISCO

INFORMATION FOR WIREMEN

On the inside page of this circular is a chart which gives practical information for supporting Wiremold Conduit and Fittings on the various kinds of wall and ceiling construction.

This chart is the result of the best thought of wiremen, contractors, architects and engineers who have standardized on the use of Wiremold for their large installations in factories, lofts, warehouses, railroad buildings, department stores, office buildings, hospitals, school buildings and the like.

Keep this chart as a guide for fastening any material to walls and ceilings.

AMERICAN WIREMOLD COMPANY.

Customer service was always a hallmark of the way The Wiremold Company did business, as this 1925 chart identifying Wiremold fittings and their uses and accompanying letter clearly demonstrate.

Consumers were catching on to Wiremold raceway, thanks to D. H.'s tireless efforts to communicate with anyone who would listen to the Wiremold story. He crafted letters for many different target audiences elaborating the benefits the Wiremold product offered distributors, electricians, industrial leaders, businessmen and even housewives.

"The men who put up electric light fixtures...never stop to think that a fixture eight feet off the floor and square in the middle of a kitchen ceiling is not only an awfully hard place to connect an electric iron, but puts all your work in the shadow when you stand before the stove or sink. Do they?"

Rapport established, the consummate salesman moved quickly to capitalize with this simple, yet effective payoff: "I think the man who invented Wiremold must have done something like that and been sorry about it afterward," he went on, "for Wiremold has made so many badly lighted kitchens nicer to work in and it is so easy and inexpensive to put in."

His frustrations aside, D. H. knew instinctively where the future of his company lay. Most immediately, it would profit from the retrofitting of domestic America, where the majority of those homes that were electrified still had an astonishing shortage of adequate outlets.

In the not too distant future, its success would pivot on the development of additional wiring systems as the nation's 6,500 power companies continued to grow and expand. 1921 was the first year since 1882 when there was a decrease in the output of electricity nationwide.

In this photograph, staff in The American Wiremold Company's office at 252 Asylum benefit from their own product after Wiremold 500 raceway is installed to carry power to new overhead lighting fixtures. The infamous "shadow in the sink," opposite, vanished once homes were retrofitted with Wiremold products.

It was an aberration. In 1922 electrical output would once again increase and, excepting the deep depression years of 1930 to 1932, it would continue to increase every year.

It was 1923, the year when America's first neon tube advertising sign was installed in New York, before Wiremold raceway finally showed a profit. Nearly a quarter of a century after he had taken control of the rickety enterprise headed by C. D. Richmondt, D. H.'s company was finally achieving its destiny, unqualified leadership in the industry, "although we didn't know it then," he would note many years later. Now well into his 40s, D. H. was far too busy doing everything he could to capitalize upon the growing success of Wiremold raceway and position the company for the future.

It would be years before The Wiremold Company stopped producing loom. And in the years to come, for various reasons, it would digress into other product lines.

Through the years, The Wiremold Company's catalogue has reflected the company's enduring relationship with its distributors. Clockwise from top left, catalogues from the 1920s, the 1950s, 2000 and the 1970s.

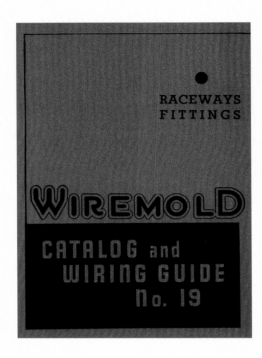

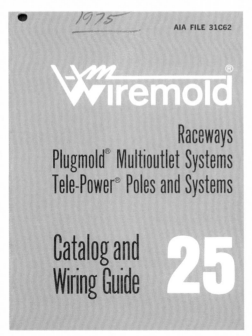

But it would eventually retreat to its core product. And that product, everyone now realized, was Wiremold raceway. ■

The company, ready to capitalize on this long-awaited success, had already been manufacturing and marketing a second Wiremold metal raceway since 1921. As its number implied, the 700 series offered slightly increased wiring capacity.

By 1924, the company had introduced the first of many enhancements that would follow, the Wiremold Bracket Convenience Outlet. Promoted in advertisements as a "safe, permanent, neat little fitting that may be attached to any wall fixture" the convenience outlet provided electrical connections for "your vacuum cleaner, iron, lamp, heater or any electrical appliance without having to remove the shades and unscrew bulbs from fixtures." It caught on right away.

But expansion of the product line was only one of D. H.'s concerns. He believed there were significant problems with the entire industry and that until they were addressed, all companies and consumers would suffer. Hoping to affect change, he became increasingly committed to the overall development of the industry.

When the old electrical manufacturers council dissolved in 1926 and was reorganized as the National Electrical Manufacturers Association (NEMA), Gerard Swope, president of the General Electric Company, was elected its first president and D. H. became vice president of the supply division.

The goals of NEMA were to enhance the strength and prestige of the electrical manufacturing industry, build stronger relationships with the government and other national organizations, increase the interest and involvement of leaders in the industry and simplify permitting and regulatory procedures.

It was for this last issue that D. H. felt a genuine passion. Increasingly, he was becoming an outspoken advocate for industry standardization. The subject would occupy him well into the next decade.

"Anything which tends toward standardization of products is of the greatest assistance in the solution of manufacturing problems," he wrote in the premier issue of the *NEMA Bulletin*, "the elimination, for example, of a condition which makes it necessary for a manufacturer

D. H. Murphy (second from right) is shown in Atlantic City in 1924 with (left to right) a United Laboratories representative, company secretary Bill Ball, Mrs. Murphy and an electrical inspector. D. H. Murphy was a leader in the electrical manufacturing industry and throughout his career worked to improve wiring standards.

Wiremold raceway was used by the Bethlehem Shipbuilding Company of San Francisco to wire the cabins of the steamships *Sierra* and *Ventura*, right, for the Oceanic Steamship Company. The fictional character known as the Wiremold Business Builder makes an early appearance in his familiar WBB bowler hat on the cover of the *Electrical Record*. This ongoing promotional strategy leant its name to a new publication for distributors: the *Wiremold Business Builder* that was introduced in 1928.

to furnish the same device in a slightly different form in different parts of the country…. Any device which is safe in Portland, Maine, should be safe in Portland, Oregon."

Toward that end, he proposed the creation of a national association for electrical inspectors, similar to NEMA, arguing that it "would be a great power for good." ■

Early in 1926, D. H. dropped the word "American" from the company name. Two years later, The Wiremold Company would introduce its third Wiremold raceway — the 1000 series.

Everywhere, now, electricity was making news. In 1928, the largest turbine installation in the world was placed in service by the United Electric Light & Power Co. at its Hell Gate Station in New York. Underwater lighting was introduced. And General Electric announced it had produced a record 3,600,000 volts of artificial lightning at its facility in Pittsfield, Massachusetts. Scarcely a year later, it eclipsed that record by generating a 5,000,000-volt lightning flash.

Fewer than a quarter of American homes had been wired for electricity when the century's third decade began. As it closed, the number of electrified homes had grown to 65 percent.

Not surprisingly, the history of The Wiremold Company seemed to parallel this trend. To capitalize on the rapidly expanding potential of American electrification, the company introduced a new publication. Published monthly for contractors and dealers, the

Wiremold Business Builder was packed with tips about how to grow business using Wiremold products. "Every 'WBB' now has the key to more business, better business and bigger profits on every job," promised a front-page headline on the second edition, in October 1928. The company had entered the 1920s reeling from one of the worst mistakes it would ever make. It ended the decade with a growing product line as it moved once more.

This time, The Wiremold Company took up residence along Woodlawn Street in West Hartford. The new building offered 39,300 square feet of space and was conveniently located adjacent to a rail line. It would remain there for the rest of the century, expanding the plant building by building to accommodate growth.

The growing importance of Wiremold raceway led the company to change its name to The Wiremold Company in 1926. The company built a new factory on Woodlawn Street in West Hartford, where the company moved in 1929.

chapter five

The Cleanest Windows In Town

"The success of our business depends upon repeat orders from the same customers. Keep them coming by giving them a perfect product."

— D. Hayes Murphy in a letter to employees, 1937

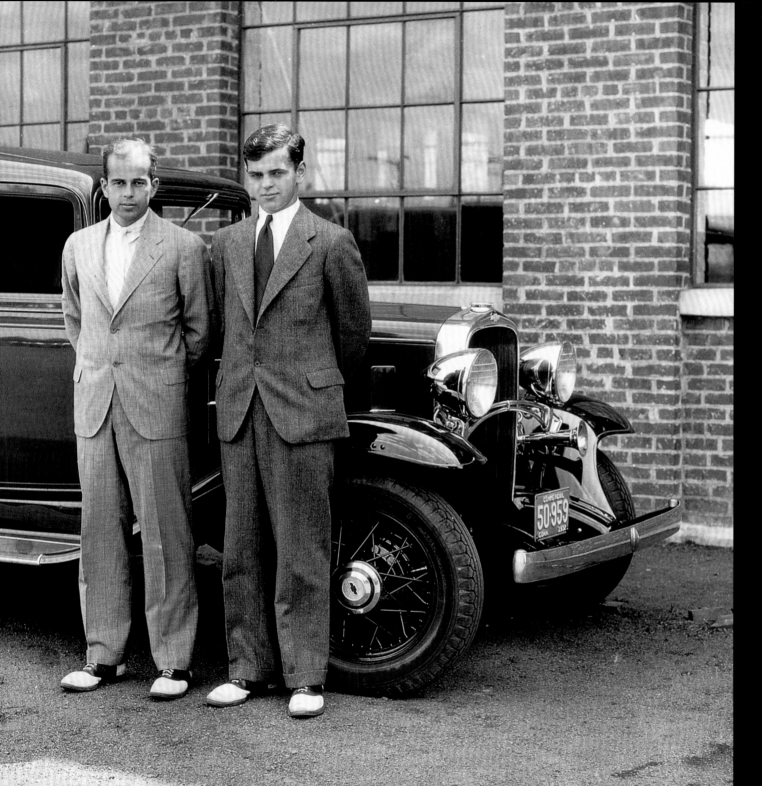

John and Robert Murphy (left to right) pose in front of a sales vehicle at the company's West Hartford plant in June 1932. D. H.'s sons worked during the summers of their college years as salesmen. John would become a full-time employee two years later. Robert would join him in 1936. Though the Depression years hit every business hard, The Wiremold Company would continue expanding and innovating throughout the leanest years.

The Wiremold Company had no sooner taken up residence in West Hartford than the economy imploded. On Thursday, October 24, 1929, the nation's overinflated stock market collapsed.

Tempered by the lessons of 1917 and 1919, The Wiremold Company was, arguably, in a better position to withstand the economic setbacks of the Great Depression that followed. Indeed, while many other businesses were failing, it managed to hold its own. Before the single worst economic decade of the twentieth century had finally worn itself out on the eve of war, the company would actually expand its physical plant twice.

Bank deposit registers for the years of the Great Depression reflect a period during which the company often had so little business that production employees were put to work washing windows.

It would, however, be a mistake to suggest the company did well during these difficult times. What growth it experienced was as much a consequence of D. H. Murphy's tenacity and gritty devotion to both the business he had built and the people who worked for him. He was resolute that The Wiremold Company would not fail. ■

Still, D. H. could not have imagined how challenging the next 10 years would be. From October 1929 until Franklin Roosevelt's inauguration in March 1933 — an unbelievable span of 40 months — the economy slid continuously downhill. American industrial output dropped by more than 50 percent with the index of industrial production free-falling from 64 to an all-time low of 56 in the three-month period before Roosevelt took office.

As panic gripped the nation and Americans withdrew their savings from bank after bank, the nation's financial industry went in the tank. Between 1930 and the day when Roosevelt declared "The only thing we have to fear is fear itself," over 5,500 banks, with total deposits of more than $3 billion failed.

The Wiremold Company dug in to ride out the economic storm. From the outset, D. H. made a decision that he would not lay off any employees unless he had no choice. It was a decision with implications that would resonate for years to come.

"Wiremold has weathered many vicissitudes...but it was during the big depression that its concern for its employees was tested to the fullest," David Bouchard reported 33 years later in *New England Electrical News*. "There was little work in the shop, sometimes virtually none, but throughout this trying period, management somehow kept most of its organization employed, often at odd jobs, many of them not even necessary."

In years to come, D. H. would look back on this time and recall, "When there was nothing else, we washed windows. We had the cleanest windows in town."

In fact, though many of the company's employees did wash windows and sweep floors through the darkest hours of the Great Depression, it was not because the man who had built the company was sitting on his laurels waiting for a break in the clouds. If the world wasn't beating a path to West Hartford, he would take The Wiremold Company to the world. ■

Starting on the West Coast, in 1930, D. H. launched The Wiremold Company's "school-on-a-truck." Salesman George Gray and his son, George Jr., logged thousands

The tone of the *Wiremold Business Builder* and the ingenuity of its promotional ideas was always upbeat and positive, a voice of hope, even in these issues from September and October of 1930, when the nation's economy had been in continuous decline for more than a year. Another two and a half years would pass before the economy — buoyed by the inauguration of Franklin Roosevelt — would show any improvement.

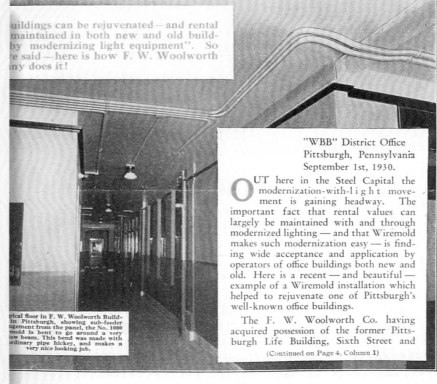

WIREMOLD BUSINESS BUILDER

THE CONTRACTOR-DEALER Sales Service—Published by THE WIREMOLD COMPANY

Number 15   HARTFORD, CONNECTICUT   September, 1930

WOOLWORTH

PITTSBURGH BUILDING

USES WIREMOLD TO MODERNIZE WITH **L-I-G-H-T**

buildings can be rejuvenated — and rental [values] maintained in both new and old build[ings] by modernizing light equipment". So [they've] said — here is how F. W. Woolworth [Compa]ny does it!

"WBB" District Office
Pittsburgh, Pennsylvania
September 1st, 1930.

OUT here in the Steel Capital the modernization-with-light movement is gaining headway. The important fact that rental values can largely be maintained with and through modernized lighting — and that Wiremold makes such modernization easy — is finding wide acceptance and application by operators of office buildings both new and old. Here is a recent — and beautiful — example of a Wiremold installation which helped to rejuvenate one of Pittsburgh's well-known office buildings.

The F. W. Woolworth Co. having acquired possession of the former Pittsburgh Life Building, Sixth Street and

(Continued on Page 4, Column 1)

[Typi]cal floor in F. W. Woolworth Build[ing] in Pittsburgh, showing sub-feeder [arr]angement from the panel, the No. 1000 [Wire]mold is bent to go around a very [l]ow beam. This bend was made with [an o]rdinary pipe hickey, and makes a very nice looking job.

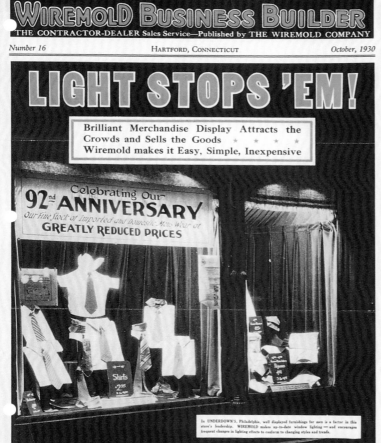

WIREMOLD BUSINESS BUILDER

THE CONTRACTOR-DEALER Sales Service—Published by THE WIREMOLD COMPANY

Number 16   HARTFORD, CONNECTICUT   October, 1930

LIGHT STOPS 'EM!

Brilliant Merchandise Display Attracts the Crowds and Sells the Goods ✶ ✶ ✶ Wiremold makes it Easy, Simple, Inexpensive

Celebrating Our **92ⁿᵈ ANNIVERSARY**
Our Fine Stock of Imported and Domestic Mens Wear at
GREATLY REDUCED PRICES

In UNDERDOWN'S, Philadelphia, well displayed furnishings for men is a factor in this store's leadership. WIREMOLD makes up-to-date window lighting — and encourages frequent changes in lighting effects to conform to changing styles and trends.

A MONEY-MAKING WINDOW-LIGHTING PROMOTION PLAN FOR WBB'S—See Page 4

of miles, from Washington to southern California, in the company's traveling advertisement. Equipped with samples of all of the company's products, as well as competitors' products, the Grays could not only explain why The Wiremold Company's products made sense, but also immediately address questions and concerns with actual demonstrations.

Though tough times meant the Grays weren't always successful selling their products, they were generally met with a warm reception. "School-on-a-truck" amounted to a free education for many electrical contractors who hadn't had an opportunity to discover the new generation of products The Wiremold Company was creating. And the program was sufficiently successful to warrant the expansion of the fleet in 1931 with a second vehicle to travel the highways and back roads of the South.

Though the business generated by this novel sales tool could not mitigate the numbing effects of the Depression, it helped. When America finally pulled itself out of this economic hole, D. H. knew, there would be a tremendous demand for building materials, and he wanted The Wiremold Company to be there with the best. So he couldn't afford to let the company's research and development efforts languish.

In 1932, a full year before Roosevelt's inauguration signaled the first upturn in the Depression, the company introduced its next new product. The largest raceway yet produced, Wiremold 1100 not only had a wiring capacity more than double that of Wiremold 500, but also was designed to accommodate multiple plug receptacles along an extended installation.

A year later, the introduction of Wiremold 200, a single-channel raceway designed for low-density wiring, made abundantly clear the company's intention to produce products that would meet all of the rapidly evolving needs of the marketplace. By now, the Wiremold line was nearly 20 years old and included five distinct raceways. ■

As the company struggled to make progress, even D. H. Murphy's sons, John and Robert, were recruited as "missionary salesmen," taking the company's message on the road during the summers of the early 1930s as they were completing their educations at Dartmouth College. Their expeditions took them on long circuits throughout New England and as far west as Chicago.

John and Robert had, in fact, been introduced to their father's business "as soon as we could walk," remembers Robert. It was a foregone conclusion that they would go to work for The Wiremold Company as soon as they had completed college, and D. H. was firm that they should know the company inside and out before taking on significant managerial roles.

By the time John, the elder, actually came on board full time in 1934, the company was in the process of its first physical plant expansion since the move to West Hartford, including a new manufacturing building (known alternately as Unit 2 or Building 2). During his first seven years with the company, John worked in every factory and office department in the company,

The company's advertising strategy consistently focused on Wiremold not simply as an inexpensive way to retrofit buildings, but as a means to profitability for various target audiences, including landlords.

NEMA
ANNUAL DINNER
1938
WALDORF-ASTORIA HOTEL   FEBRUARY 10, 1938

DRUCKER-HILBERT CO
NEW YORK
8238B

D. H. Murphy had been president of the National Electrical Manufacturers Association (NEMA) for a year when the organization held its annual dinner at the Waldorf-Astoria Hotel in New York City early in 1938.

thoroughly learning what made it tick. Robert, who gravitated to the mechanical side of the business, joined his brother in 1936, starting as a clerk in the factory office. He would remain in that role for five years before working for a time as a machine operator and then as a draftsman. In time, the brothers would make their own significant marks upon The Wiremold Company. But for now they were learning about the company in some of the worst of times. And what they learned, not only about their industry, but about survival, would serve them well for the duration of their careers. ■

As The Wiremold Company completed the first half of Unit 2 and a new two-story office building in 1935, it seemed like more than merely a physical plant expansion. To many it seemed a powerful symbol of hope. The economy was actually showing modest signs of life. America held its breath, hoping that a recovery might be coming.

The company expanded its product line yet again the following year, introducing the 1500 overfloor raceway. Sometimes referred to as "Pancake," 1500 was a flat raceway designed to carry wiring from walls across floors to installations in the center of rooms, such as desks or equipment tables, with minimal obstruction. The introduction of 1500 was followed by the completion of the second half of Unit 2 in 1937. Surely now there was no question that things would get better.

But, no sooner had the new addition opened than the economy took another dive. Over a miserable 13-month span that stretched from May 1937 all the way through June 1938, industrial output once again sagged, dropping by more than 32 percent. Like a stubborn fever, the country's jobless rate rose to a high of 20 percent, second only to the high of 25 percent reached during the first three years of the decade.

The Wiremold Company was undaunted. As if defiantly waving a totem at the ruthless gods of economics, the company proceeded with plans to introduce another new product, 2100 metal raceway. In the seven worst economic years the United States had ever experienced, The Wiremold Company had introduced four new products and expanded its physical plant twice.

Also in 1937, D. H. became president of NEMA. In this role, he was in an ideal position to promote The National Adequate Wiring Program, an educational effort to encourage homeowners to press the industry for more adequate wiring. Launched in 1936, the program was a big hit with contractors, who saw it as a banner for the future of the industry; a rallying cry that would not only drive the electrification of America, but could also revive the economy.

Still, as 1937 wound down, employees faced another

holiday season dampened by the intractable Depression. In the profit-sharing letter employees received on Christmas Eve that year, D. H. gave voice to the optimism that had kept The Wiremold Company afloat throughout those difficult times.

"Although business has fallen off during the last half of the year we are fortunate in having made good progress during the first half and 1937 will go down in our records as a profitable year," he wrote. "Profits made were plowed back into the business in the form of new

The company's West Hartford plant was expanded in 1935 and again in 1937. Above, D. H. Murphy is pictured in his office in the new administrative building, completed in 1935. The roll-top desk he bought in 1911 is visible behind him.

Fluorescent lighting, enormously popular from the moment it was introduced in 1938, was a welcome shot in the arm for The Wiremold Company, which created a new lighting division in September 1939.

combination of circumstances which makes possible this large distribution of Preferred Stock among those who have worked together so faithfully to make The Wiremold Company what it is and we are confident that the increased holdings in the hands of each will add interest to our work and thereby increase the value of the stock which each of us owns.

"With best wishes for a Merry Christmas." ■

The economy would finally start to reverse itself in the middle of 1938. And though it would take a war to actually drag the nation out of the grip of Depression and focus the national psyche on a shared enemy that was tangible, something else would first raise the cloud of darkness off the electrical industry.

The light of hope for the industry was precisely that — a light called fluorescence. Introduced in 1938 to great fanfare, fluorescent lighting quickly became enormously popular, both because of its novelty and because it was economical. It was so popular, in fact, that by September of 1939, The Wiremold Company had announced a new lighting division headed by Perry S. Strang. The lighting division was soon doing a tremendous volume of business, employing a powerful advertising campaign that targeted industrial, commercial and residential customers, explaining how the company's fluorescent lighting products could help address their needs.

For D. H. Murphy, the new lighting business paralleled his personal commitment to the National Adequate

building, new machinery, new tools, new equipment, inventories, moving expense and development expense on new products.

"This means that our gain is not in cash but in the above mentioned items and it has been decided to distribute a substantial percentage of this gain as extra pay in the form of Preferred Stock to our fellow-workers who have been with the company two years or more. The amount of stock which each one will receive is 21 percent of earnings.... We are thankful for the

Wiring Program. In his role as president of NEMA, he published a major article in *New England Electrical News* in which he aggressively addressed the need for "modern electrical facilities" and urged his readers to relentlessly press the industry for more adequate wiring. He proposed "plenty of branch circuits with adequate copper in those circuits and an ample number of convenience and lighting outlets with proper switch control in home wiring situations." Housing starts had been down for nearly a decade. He knew the corner was just ahead.

Before The Wiremold Company would be able to take full advantage of that burgeoning market, however, the company would need to become unionized. Repeatedly, throughout the decade, the company's sales force had encountered resistance to its products in major markets such as New York City, where trade unions had grown increasingly strong.

In 1939, 25 years after The Wiremold Company's first experiment with profit sharing began, it signed its first contract with the International Brotherhood of Electrical Workers (IBEW). The event followed acrimonious struggles and strikes in the steel and automotive industries. But The Wiremold Company's employees had never asked to join a union. Nor did they ever initiate a work action of any sort against the company once the union was in place.

Twenty-four years later, David Bouchard reflected on this in the article he wrote for *New England Electrical News*. Bouchard attributed the company's successful relationship with its employees, as well as the company's remarkable safety record, to the loyalty D. H. Murphy

engendered when he refused to lay off his workers during the darkest days of the Depression. "My father was never buddy-buddy with his employees," recalls John Murphy. "He was 'the boss!' But he cared about people and he believed in fair treatment."

From a marketing standpoint, unionization worked, opening doors almost overnight in major markets. Before the 1930s ended, a sixth new Wiremold product, 3000 raceway, was added to the rapidly growing product line. The first two-piece raceway (base and cover installed in the field), 3000 was a perimeter product, designed to provide ample capacity while fulfilling the aesthetic demands of baseboard. And by 1940, the company was announcing plans for a new $100,000 expansion of the plant only three years after completing Unit 2.

The worst decade in the history of the nation's economy had begun with salesmen on the road, demonstrating The Wiremold Company's products, and so it ended. Riding the growing crest of the new housing market he had predicted, D. H. penned another of the articles that would increasingly occupy his attention as he grew older. But this time he allowed himself the brief luxury of self-congratulation.

"More than 20 years ago," he wrote, "the founders of Wiremold started to crack a prejudiced market and successfully turned doubting Thomases into buyers." For another brief moment, not unlike the months of 1929 preceding the collapse of the economy, it must have seemed to him that The Wiremold Company finally would be able to seize the elusive gold ring.

chapter six

Wartime Solutions

"Regardless of how deeply absorbed in war we may be, this gang is definitely Wiremold-minded, and every war job problem we solve gives us a new idea for the postwar business we are planning."

— D. Hayes Murphy in a letter to distributors,

December 1944

Ceremoniously accepting the Army-Navy E Flag for The Wiremold Company are D. H. Murphy (center) and employees of The Wiremold Company, including union representative Ed Quinn (far left); Lucille Francoer (to Murphy's right), representing the company's post-Pearl Harbor employees; Mary Cote, representing the company's pre-Pearl Harbor employees; and Henry Martoccio (far right), chairman of the company's War Production Drive Committee.

When the Great Depression finally bottomed out, most of the impediments The Wiremold Company would have to survive were behind it: the early years before D. H. finally consolidated full control of the company…the shortages of World War I…the near-disastrous loom problem that came on the heels of the company's move to Hartford.

But the company would still need to navigate the economic circumstances during the greatest war in the history of the world before it would finally discover real prosperity in the burgeoning postwar domestic market. Short of the unethical and the illegal, there wasn't much D. H. Murphy wouldn't do to keep the blood flowing through his company's veins.

When Anton Chernak came to The Wiremold

D. H. Murphy poses for a publicity photo with a group of employees and executives circa 1941. In the back row are Robert Murphy (far left) and John Murphy (second from right).

Company in 1941, it was to sell an idea. He was a Russian immigrant, a natural genius with a flair for machines. After arriving in the United States, he had spent some time working in the textile industry in Rhode Island. But his heart was not in textiles. He professed to D. H. Murphy that he could revolutionize the prevailing flexible-air-duct technology.

The existing technology made possible production of air ducts on a lathe in lengths up to 20 feet. If Chernak's idea worked as he said it would, The Wiremold Company could produce flexible air duct economically in greater lengths. Not that any of this would have necessarily made a difference to the company under ordinary circumstances. Indeed, air duct had practically nothing to do with the company's core products. But the Army Air Corps needed a large diameter flexible duct for preheating aircraft engines in Arctic environments. And in 1941, D. H. Murphy saw that military contracts might well be the buoy that could keep his company afloat over the next few years.

So he brought Chernak on board. Since the idea for the duct existed only in Chernak's mind, D. H. assigned Robert H. Francis, a company engineer, to help him turn the idea into reality. A new department was formed to explore the manufacture of air ducts.

Though the company would commence 1942 by introducing yet another Wiremold raceway, number 2600, as well as the new Air Duct Division, it already was modifying many of its standard products for military use. Indeed, the urgent need to quickly convert buildings and

factories across the country from peacetime to war applications would test Wiremold raceway's convenient and
flexible application claims as domestic use never had.

The products, in turn, would live up to their reputation. By war's end, practically every one of the company's
products was in extensive use in thousands of military
installations. Loom, for example, was used for wiring
every kind of military vehicle imaginable, from tanks to
trucks to airplanes.

But as the company stepped up to its obligation to
help with the war effort, the core product line was soon
augmented by a remarkable array of other products for
all sorts of military applications. Largest among these
products was webbing, a form of flattened loom used by
the U.S. Army for parachute harnesses and as strapping
to secure supplies delivered by parachute. Converting
the loom machines in its textiles division to web production, the company earned a lucrative contract from the
Quartermaster Corps and soon was generating more
than 1.6 millions yards of webbing per month.

At the same time, the company's stamping departments were busy fulfilling an array of sub-contracts, as
well. The war had an unquenchable appetite for materiél:
306,133 Pratt & Whitney engine deflector parts per
month, nearly 50,000 Grumman intercooler parts, some
7,000 fuselage parts for Sikorsky, nearly a third of a mile
of heater tubing for aircraft engines, upwards of half a
million navy rocket shipping containers, 138,000 shell
shipping containers and more than 5,000 glider parts.
Month after month after month, The Wiremold

Company kept producing.

The crushing bleakness of the Depression was
replaced with a sense of purpose. Certainly the country
was embroiled in a horrendous conflict, and young
Americans were dying in unprecedented numbers in
places most Americans had never heard of. But there was
work, and there was a cause to which one could devote
oneself.

Where The Wiremold Company had struggled to
find enough business during most of the 1930s, it was
confronted now with a whole new set of challenges. How
could the company keep its domestic business alive and
reliably predict that magical moment when the war

An inspector looks at a
batch of the more than
1.6 million yards of
webbing The Wiremold
Company produced for
the U.S. Army during
World War II. The com-
pany's loom machines,
once reconfigured, were
ideal for making the
product, used for
parachute harnesses
and similar strapping.

The Hartford Area Labor-Management War Manpower Committee, below, faced chronic shortages of workers. John Murphy (standing, second from right) was one of the several business leaders and union officials on the committee. Right, office workers take a break in the newly refurbished ladies' lounge in 1943.

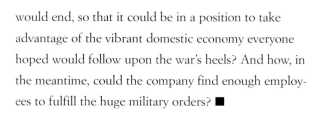

would end, so that it could be in a position to take advantage of the vibrant domestic economy everyone hoped would follow upon the war's heels? And how, in the meantime, could the company find enough employees to fulfill the huge military orders? ■

Rosie the Riveter wasn't only an inhabitant of U.S. Navy shipyards. By the spring of 1943, a woman who could operate a sewing machine had an excellent chance of finding work at The Wiremold Company as military contracts expanded to include parachute cords, tent flaps, knapsacks and belts. Women working at the company participated in exhibitions aimed at attracting more of their ranks for what the *Hartford Courant* described as "thousands of positions available to women in hundreds of Greater Hartford war plants and business concerns."

During those years, office staff helped fill gaps on the factory floor. After working all day in the office, they moved onto the factory floor for the second shift. John Murphy, general manager at the time, worked in the Air Duct Division on the second shift. Edith Wootton (who later married John Murphy), spent days in the advertising department and the second shift operating a four-needle sewing machine.

The company, in turn, introduced a flurry of innovative employee benefits. In March, for instance, a new pension plan, with significant advantages for employees, was announced. The purpose of the plan was to ensure all employees a retirement income which, together with Social Security, would amount to a reasonable retirement income. The new plan required only two years of service for eligibility and also featured a death benefit.

Since the previous year, the company had sponsored a large "Victory Garden" on the grounds just outside the plant. Now, as a further response to the problem of food rationing, the company opened a new cafeteria on May 7, 1943. At its formal dedication, several significant groups of employees were honored. They included the newly formed 10 Year Club, with 75 members who had devoted at least a decade of their careers to The

Wiremold Company; the company's popular basketball team, league champions the previous winter; and The Pittsburgh Platoon, the five men who came east from Pittsburgh with the company in 1919 — Treasurer Louis S. Zahronsky, Secretary William D. Ball, Production Manager James M. Foley, head of the textile division Charles Rutherford and D. H.

At the same time that these innovations were being introduced, a wide range of techniques were also being used by the company's personnel department to help keep up the morale of a workforce with many relatives

Edith Wootton Murphy played a key role in the creation and execution of The Wiremold Company's advertising and marketing communications over her career and manned a sewing machine on the second shift during the war.

Victory Garden

As early as 1929, employees, with the consent of The Wiremold Company, had been planting small gardens that they tended before and after work. During World War II, the company expanded this program, plowing up a large piece of land on the opposite side of the plant from the railroad tracks. The company fertilized and treated the soil and turned it over to 15 employees, who were delighted at the opportunity to augment their home larder.

By 1943, as other employees saw how productive the garden was, so many workers applied for the 50 1,500-square-foot garden plots that the company had to establish an employee committee to evaluate the applications. First priority went to employees who didn't have room for a garden at home. Participants also had to pledge to can at least 50 quarts of vegetables and 35 quarts of fruit, or enough to make their families self-sustaining until the 1944 harvest. D. H. Murphy, who vigorously endorsed the project, added a third condition — straight rows.

When employees raised more produce than they could use, the new company cafeteria purchased the excess. Workers came early, stayed late and often devoted their lunch hours to tending their crops. Some, like Charles Hall, foreman of the punch press department, experimented with orna-

Good produce and employee goodwill grew in The Wiremold Company's enormously productive Victory Garden.

mental flowers. Others, like Tobe Jones, tried their hand at growing peanuts. The Victory Garden was not work; it was a symbol of hope in troubled times.

Responding to employees' enthusiasm, the company introduced livestock to the experiment in social agriculture. Pig pens and chicken coops were built and several hives of bees were installed to ensure crop pollination.

The chickens were a success. Starting with a flock of 300 birds, the company was soon producing a wealth of eggs and spring fryers that augmented the rapidly expanding and increasingly nutritious menu in the cafeteria. The pigs, on the other hand, smelled so bad and were so difficult to transport to a butcher that the company discontinued its pork "subsidiary" after a year.

"From now on, our hats are off to the pig farmers of America," reported the company newsletter *Wiremold Outlet*, "and we will do what we can to help them get electrified (with Wiremold, of course) — but as far as encroaching on their sacrosanct business of pork production, we retire while we can...."

Bonded not only by their work, but also by their shared commitment to victory overseas, employees raised both their spirits and tons of fresh produce. In 1945, The Wiremold Company was awarded top honors from the National War Gardens Committee for its successful Victory Garden program.

engaged in armed conflict abroad. Posters displayed throughout the company urged workers to do their part for the war effort. Letters from former employees in the service were displayed on bulletin boards along with absentee charts to remind workers that every day of work they missed helped the enemy. Veterans who returned from the battlefront were frequently invited to The Wiremold Company to speak to workers. And patriotic music was broadcast over the company's loudspeaker system.

Like Americans everywhere, the company's employees were also encouraged to purchase war bonds. By the summer of 1943, the company had been awarded a U.S. Treasury flag, indicating that 98 percent of employees were participating in the war bond payroll deduction plan. Employees who purchased bonds in excess of the payroll deduction plan were treated to rides in a Jeep, the popular new military vehicle that, according to the *Hartford Courant* was "the new American ambition." ■

Though these innovations helped to address the chronic shortage of workers that The Wiremold Company was now plagued with, company management was by no means ignoring the long-term concern of growing the company for its future beyond the war.

Robert Murphy, elected assistant treasurer in 1941, and Safety Director George Pieper, for instance, became charter members of the Hartford Industrial Safety Council and took a 20-week course taught by H. W. Heinrich, of the Travelers Insurance Co., one of the

Thanks to innovations pioneered by Robert Murphy, The Wiremold Company has long enjoyed an enviable safety record. In the 1940s, punch presses like this one were equipped with warning flags to signal any mechanical defects in the clutch, pressure switches that would shut the machine off when air pressure dropped too low and interlocking relays that made it impossible to double trip the press. By using several redundant safety devices, the company was able to greatly reduce the likelihood of accidents.

leading experts on industrial safety in the nation. The Wiremold Company had a longstanding reputation for safety, and management was determined to ensure the company remained a safe place to work, especially with many new employees handling jobs with which the company had little experience.

For Robert Murphy, safety was more than merely a policy question. He realized that many safety problems could be solved with technology, and with his guidance, the company pioneered many safety devices. One that is still talked about on the shop floor involved punch press machines.

Punch presses were among the most dangerous machines on the manufacturing floor. Workers were warned not to put their hands inside when the press was in operation. But Robert analyzed the operation of the machines from an engineering standpoint and realized that he could design a way to absolutely prevent workers from putting their hands into the machines. He invented a set of sleeves, linked to the machines' operating switches, that allowed workers to do their jobs, but made it impossible for the workers to put themselves in jeopardy.

The solution worked, although it created controversy for a time. Visitors to the plant sometimes got the wrong impression that the company was chaining its employees to their workstations.

"My father understood the importance of safety and we were never turned down for anything we requested to make the plant safer," Robert recalls. "What we realized, early on, was that safe working conditions didn't slow things down. Workers who knew that the machines were safe felt more comfortable working at capacity. Safety is good business."

At the same time, D. H. was hard at work shoring up relations with electrical distributors that would serve the company well when the war was over. In selecting materials for military construction projects, the War Production Board's construction branch was adhering closely to a list of electrical materials that did not include The Wiremold Company's products.

The company put together a strong case that not only argued the reliability of its products, but also highlighted that, for many applications, Wiremold raceway was the best possible solution. When the materials were presented to leaders of the Board's construction branch, the policy was rewritten. In a publication for its members, the Eastern Electrical Wholesalers Association printed the letter from D. H. explaining the steps the company had taken to address the problem. Prefacing the letter, the Association wrote, "We herewith quote a fine...letter from The Wiremold Company to show that not only has this fine company their Distributors' interest at heart, but, *unlike others*, they are doing something about it." ∎

On August 4, 1943, The Wiremold Company was presented the Army-Navy E Flag, an honor awarded to companies for excellence in production of war matériel. Before the war ended, the company would earn three E Stars to embellish the flag. Relaxing for a few hours from the relentless pace of war production, the company held

the E Flag, the webbing contracts, the primary piece of business on which the company had depended for the past two years, were completed. And there was nothing immediately in sight to replace those contracts.

When D. H. sent his employees their final profit-sharing letter of the year, on December 17, he reported that 1943 had been a little better than 1942. A bonus was earned during the first nine months of the year, before the loss of the webbing contracts. But no one could be sure what lay ahead. At the company's annual meeting the following February, D. H. reported, "As we are essentially a peacetime outfit, and this was another war year, we have had our problems."

At a meeting of company supervisors, foremen and union officials in 1944, D. H. reported on efforts by the governor of Connecticut and the mayor of Hartford to address the issue of manpower shortages, but said more efficient use of current manpower was the only real solution. "Our having earned no bonus in the first three months of the year certainly is bad news," he said, but he added that the company was "hard at work trying to build up our volume of business in the hope of replacing [the webbing contract]. Some progress is being made."

Reflecting the ongoing positive relations between the IBEW and the company, Edward T. Quinn, president of Local 1040, said, "If the company operates for three months and doesn't make a profit, it's bad for all of us,

an open house to celebrate.

In fact, there was little to celebrate. Hard work and awards aside, the war years were not good years for The Wiremold Company. The military contracts kept employees busy with important work they could take pride in. But for four more years, war put the company's growth on hold.

A month after The Wiremold Company was awarded

Airplane parts that went into Vought Corsairs, below, were just part of the war materiél The Wiremold Company produced earning the the Army-Navy E Award on August 4, 1943, opposite. John Murphy, accompanied by a trio of company officials, rides shotgun in a Jeep in June 1943. Rides in the vehicle — developed for the war — were given as a reward for the company's record participation in war bond drives.

because we get no bonus. Let's try to stop the waste."

Throughout the year, the company picked up additional military contracts, culminating, just before Christmas, with the company's second E award for production excellence.

But more and more, D. H. focused his attention on what would happen when the war was over. The Allied forces were making progress. In December, 1944, he sent his distributors a message of faith. "Christmas finds us embarked on our fourth year of war," he wrote, "and

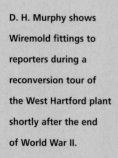

D. H. Murphy shows Wiremold fittings to reporters during a reconversion tour of the West Hartford plant shortly after the end of World War II.

An Operator is needed here

in September, 1945, $35 billion in war contracts were abruptly canceled.

It was a stinging blow to the economy, but not an unexpected one. As early as 1943, the War Production Board had begun planning for the war's end, and the result was a conversion back to peacetime economics that happened much more quickly than after World War I. By October, 1945, the War Production Board had been replaced with a Civilian Production Administration. And by the end of November, peacetime employment had bounced back to the end-of-the-war total and 93 percent of war plants across the country had been reconverted.

The Wiremold Company was ahead of the curve. The war ended before the U.S. Army Air Corps got to use the flexible duct that Anton Chernak and Robert Francis began developing in 1941. To keep the ongoing development of duct from interfering with the war effort, the company farmed out the project to Sansone Tech., a Hartford tool and die company, where Chernak and Francis's ideas were refined by William Rejeski and Charles Frederickson. By 1946, when The Wiremold Company brought the production back in-house (and Rejeski with it), the Sansone team had turned Chernak's idea into an auto heater hose concept that would soon dominate the market.

"Thanks to the work Chernak and Francis had done, we had a head start, and we were in a position to jump into the auto industry just as soon as it recovered after the war," recalls Paul Roedel, who joined the company

we are more determined than ever that this must not happen again...

"We are always looking forward to that day after Victory when Wiremold production will again have the green light. On that day, you may expect us to knock on your door seeking an opportunity to demonstrate that our wartime experience has prepared us to serve you better than ever before." ■

D. H.'s prognostications came to fruition, but not without some pain. After Germany surrendered, cutbacks in military spending began. And when Japan surrendered

that same year and would head the Air Duct Division.

"Our duct could be tapered and worked into shapes, and it could be made using various fabrics for different applications," Roedel says. "It had distinct advantages over the vulcanized duct that was then in use.... Our entry into the automobile industry was in production of heater and defroster hose, but our share of the market just grew from there." By October, scarcely a month after the Japanese surrender, the company had already shipped its first order of the new product.

Piece by piece, contract by contract, the company was making steady progress in D. H. Murphy's postwar plan for production diversification. By the following summer, the company had consolidated the air duct business into a new production division for the manufacture of special parts, including air ducts for heating and ventilation, components for the Royal Typewriter Co., and parts for metal chairs and tables, all new product lines for the company.

D. H. Murphy's tireless efforts to maintain positive relations with his domestic distributors throughout the years of the war were paying off. And every employee who had gone into the service returned to work for the company after being discharged. Demand for space became so intense that the company erected a large Quonset hut west of the plant, on the space where, only a year earlier, employees had still been harvesting vegetables from the Victory Garden.

On October 31, 1946, D. H. distributed a guardedly optimistic letter to his workers. "If things work out as we

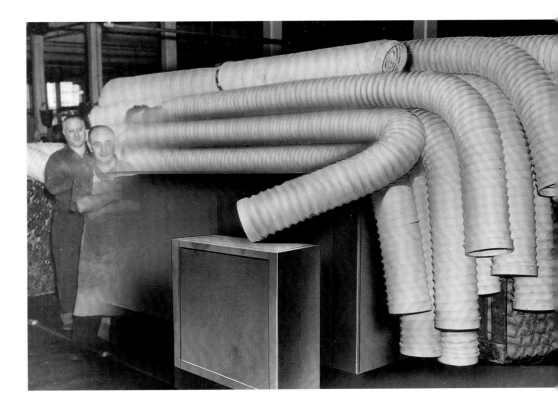

have planned them for the fourth quarter, your share is going to be more than it was during any previous quarter this year," he wrote. "And I venture the further prediction that...the year 1947 will be the best year that we have had in a long time.... Let's resolve, each and every one of us, to make the most of what looks like a golden opportunity!"

When employees received their profit-share letters on December 18, 1946, they received a bonus reflecting the best three months The Wiremold Company had enjoyed in a long time — a very long time.

By November, 1945, two months after the Japanese surrender, more than 90 percent of war plants across America had been reconverted and the first order of flexible air duct, which would become a post-war gold mine for The Wiremold Company, had already shipped.

chapter seven

"The Market is Here...Now"

"We can sell really adequate wiring

now because adequate wiring is

something that our customers —

whether they're home owners or

home builders — want, need and

are ready for."

— D. Hayes Murphy in a letter to salesmen and

distributors, 1953

House" and the *Better Homes & Gardens* 5-Star "Readers Choice Home" specified Plugmold raceway for builders. With state-of-the-art kitchen appliances, these popular homes — positively futuristic — became the templates of an enchanting, optimistically sunny new America called the suburbs.

In the summer of 1953, as the war between North and South Korea stalled in an inconclusive cease-fire that remains in place to this day, the dawn of America's new housing and construction boom was fueled by significant improvement in the production and availability of steel. And before the year was over, the two model homes were being replicated, complete with Plugmold 2000, for a new generation of young families in Detroit, San Francisco, Spokane, South Bend, Dayton, Tulsa, Cincinnati, Los Angeles, Colorado Springs, New Haven and on and on across the face of America.

The opportunities kept coming, and the company sought to take advantage of this boom. The communications department introduced a new publication, *Wiremold Electric Ideas*, in 1953. Conceived by Charles Brunelle, who headed the company's advertising agency, and Edith Wootton Murphy, it was a dynamic newsletter serving as a two-way informational conduit between the company and its customers.

The company gathered creative ideas from customers about productive ways to use the company's products. With accompanying photographs and illustrations, current examples of products in action were published and fed back to contractors, distributors and consumers in a constant stream of creative ideas. The new publication was a huge success that harked back to D. H. Murphy's first efforts to educate the customer about an innovative new product called Wiremold raceway. The newsletter also served to keep The Wiremold Company in touch with customers' needs, aiding new product development. ∎

In spring 1954, in a *Qualified Contractor* article titled "The Doctrine of Electrical Sufficiency At Last Has Caught Us," D. H. Murphy wrote, "Years from now, when we look back on this awakening of the industry to the need for adequate wiring, I think we'll find that it was the women who manage and operate our homes who were responsible for it. They have made it clear to everybody concerned that, in order to carry on their jobs, they must have at their fingertips everything in modern labor-saving appliances and devices that the electrical industry has to offer."

A product review that appeared in *Contractors' Electrical Equipment* also in early 1954 focused more industry attention on the fact that the builders of the Trade Secrets House and the 5-Star Readers Choice Home "had found from the day the model homes were thrown open to the public that Plugmold 2000 proved a major selling feature." John Murphy, executive vice president by this time, seized upon this opportunity instantly, sending letters to hundreds of builders, architects and electrical contractors, promoting Plugmold products.

John's letter was a precursor to the "Plug Plugmold"

The company launched its "Plug Plugmold" campaign, above, one of the most effective promotions in Wiremold history, in the autumn of 1954. In 1956, Plugmold 2200 was introduced, expanding the familiar Plugmold line with a wider raceway that took the place of wooden baseboard. The W. E. Parks home in Plantsville, Connecticut, opposite, was decorated for the 1960 holiday season with ample accommodations for lighting thanks to Plugmold 2200.

Waremold

At The Wiremold Company, the term "family company" has always had a meaning greater than the company's creation and ownership by the Murphy family. D. H. Murphy learned early on that one of the most effective tools he could use to find reliable new employees was the referrals of current workers. Most new employees recommended by Dad or Mom or Uncle Ray ended up making a real contribution to the company. They had an added incentive to work hard. They had a family reputation to uphold.

To this day, dozens of families are represented in the company's workforce. Some families have been employed for several generations. But aside from the Murphys, no family has ever achieved a higher profile at The Wiremold Company than the Ware family.

The Wiremold product had not yet turned a profit when Richard "Buster" Ware, who had moved to Hartford from Georgia, became the first member of his family to join the company. It was 1922.

Buster's brother, Walter, joined him in 1924. And Walter's sons — Fred, William and Richard, (Freddie, Willie and Pop to everyone who knew them) — joined the company in 1936, 1941 and 1945, respectively.

Although they distinguished themselves as employees — working in practically every area of the company's operations — their contributions to the company were greater. Willie was active in the union. And he and Freddie, both talented athletes (Fred had played on semipro baseball and basketball teams in

The Wiremold Company's basketball team, associated with the Hartford Industrial League, was the frequent champion. The 1946-1947 team members pose here to commemorate their league victory.

the 1930s) helped to make the company's teams among those most feared by competitors in the industrial leagues.

What The Wiremold Company meant for the Wares, in turn, was the kind of opportunity many African-Americans had difficulty finding in an age when segregation was still an ugly reality in America. "My husband and his family were always grateful that the company made it possible for us to live good lives and take care of our families," says Mrs. Bernice Ware, Fred's widow. "All of my children went to college, thanks to The Wiremold Company."

And while they were going to college, Ferne, now a consultant in Arizona; Gary, a training manager with Stop & Shop Supermarkets; and Bradley, an attorney with the Federal Bureau of Investigation, all worked at The Wiremold Company during the summers.

In 1984, three years after Willie's retirement and six years after Fred's, the company paid special tribute to the Wares in a *Wiremold World* article that recalled one of Willie's favorite lines: "When the Wares take over the company, it will be easy to change the 'I' to an 'A,' and then it will be called The Waremold Company."

Willie and his family had, collectively, spent more than 200 years working for The Wiremold Company. But the article also highlighted a side of the Wares' life that many of their fellow employees might have known less about — their unflagging devotion to the community.

Between them, Freddie and Willie Ware gave nearly a century of active service to the Hartford Parks and Recreation Department, building and maintaining community playgrounds, coaching youth teams, and helping make life better for Hartford youngsters.

In Hartford's North End, youngsters still play and start to develop their athletic skills at a popular Windsor Street playground named for Willie Ware. And in 1996, The Urban League named an annual program of community athletic competition after Fred.

campaign the company would launch that autumn. Targeting distributors, the campaign included a prize catalog. For every dollar in Plugmold sales distributors accounted for, they were issued a stamp worth two points. Focusing on the challenge to "Cash in on the $15 billion national rewiring market," the campaign ran from fall 1954 until July 1955.

Before "Plug Plugmold" ever got off the drawing board, however, the product had gotten another shot in the arm. In June, *House & Garden*'s 1954 "House of Ideas" was constructed on Mohawk Drive in West Hartford, practically in the company's backyard. Built by Green Manor Estates, furnished by G. Fox & Co., and designed by prominent architect Philip Ives, it featured Plugmold 2000 in the state-of-the-art kitchen, which was appointed with a refrigerator-freezer, double oven range and automatic dishwasher.

Clearly, The Wiremold Company's golden moment had finally arrived. *Nation's Business*, reporting that fall on a survey of the need for more and better wiring to accommodate America's burgeoning postwar growth, estimated that at least 30 million buildings needed rewiring. And *Wiremold Business Builder*, in turn, estimated this represented a $15 billion market opportunity and launched "Plug Plugmold" on the strength of that estimate.

"Circuits are being overloaded dangerously because householders generally do not understand the nature of electrical circuits, and the need for more of them as uses for electricity multiply," *Business Builder* told readers.

"The electrical contractor can point to this when wiring new buildings or remodeling old ones, show the vital importance of adequate wiring and explain how Plugmold provides multiple outlets safely and at a low cost." ∎

As The Wiremold Company reached the middle of the 1950s, business had never been better. And it was in this context that John and Robert Murphy at last reached the top of the company's management structure. On April 6, 1955, John was named president of the company, a title he would hold for more than 20 years, and Robert was named executive vice president.

They presided over a business at the top of its game. Though the market for loom continued to decline, the textile division nevertheless paid its way. All of the company's other products were doing phenomenal business.

Plugmold products were flying out the door. And the air duct business, a beneficiary of America's new love affair with automobiles, had one record year after another.

By now the duct business, based originally on a single kind of defroster hose, was producing air conditioner ducts, air intakes for carburetors, and parts for air distribution.

"We got off to a head start, and our place in the auto industry grew steadily," recalls Paul Roedel. "Pretty soon we were selling directly to Ford, General Motors, Chrysler and all the others. These were exciting, heady times. The auto industry showed no bounds, and we were fortunate to be able to go right along with that."

By the end of the decade, The Wiremold Company would become the largest supplier of automobile-air-duct products in the world.

Simultaneously, the company's electrical products demonstrated equally striking growth. In 1956, 2200 series products were added to both the Wiremold line and the Plugmold line. Plugmold 2200, a multi-outlet perimeter raceway designed as a baseboard, indicated the growing importance of aesthetics in new product development. That same year, *Wiremold Business Builder* reported over $14 billion would be spent in 1956 for improvement, maintenance and repairs of 11 million American homes. Over the next two years, the company continued to aggressively promote Plugmold products, with advertising campaigns in the *Saturday Evening Post* and *Time* that reached millions of readers.

Sales in 1959, propelled by the company's aggressive marketing effort, topped 1958 sales by 22 percent, and in the company's annual report, published March 25, 1960, John and Robert tipped their hats to both the advertising and to the company's sales team.

"Always with an eye to the future, we make every effort to be represented in all organizations and at all meetings that will help us to design and make better products, and then offer them for sale to the customers who need and want them," the brothers wrote. As examples of such industry involvement, they pointed to The Wiremold Company's presence at that year's Annual Wiring Sales Conference in Philadelphia; close contact with the

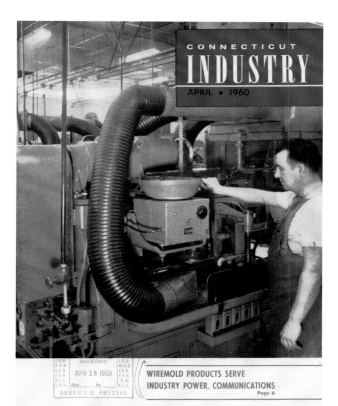

WIREMOLD PRODUCTS SERVE
INDUSTRY POWER, COMMUNICATIONS
Page 6

The Wiremold Company was featured in the April 1960 issue of *Connecticut Industry*. The cover article described the company's exemplary personnel and safety programs and other management initiatives the Murphy brothers were working on.

National Association of Home Builders, which led to research collaboration with the Massachusetts Institute of Technology on new housing components; and participation in a project concerning the design of science education buildings, sponsored by the American Institute of Physics and the American Association of Physics Teachers.

The need to solve America's wiring deficiencies and the fact that The Wiremold Company offered cost-effective solutions continued to dominate the company's marketing efforts into the 1960s. The National Adequate

Wiring Program, in fact, would remain a cause for NEMA until 1965.

In February 1959, John and Robert had introduced a program to reduce the company's costs through "streamlining operations and flow of materials, cleaning out obsolete items, etc." They hoped that through consolidation and aggressive clean-up The Wiremold Company would save enough room so that further building construction wouldn't be necessary.

In late March, little more than a month after the program was launched, the brothers reported that, "we now have great hopes that when the present program of housecleaning is completed, we will have sufficient room for at least two or three years." A possible exception, they noted, was the main office, which had not been increased in size since it was built in 1935.

As it turned out, John and Robert Murphy's clean-up effort had more than achieved its goals. The company was able to avoid constructing new administrative space for another eight years. They would have no way of knowing just how important such consolidation would become before the company reached the turn of the century.

It wasn't until 1965 that a new 57,000-square-foot factory addition, Building 4, increased the company's manufacturing space by some 40 percent. Nearly a quarter of a century had passed with no significant change in the company's physical plant. But riding the crest of the 1950s housing boom and the automotive craze that accompanied it, the company had prospered.During

1963 and 1964, the company had been forced to staff up significantly, adding 100 new employees. And the Murphys fully expected to add more based on the continued growth of electrical and air duct product lines. The new building made a bold statement: These were good times.

"This new building is required to expand our production facilities and make them more efficient and more comfortable," said President John Murphy at the opening in 1965. "This new expansion plainly demonstrates that The Wiremold Company plans to remain right here. This area has been good to us ever since we first opened our doors nearly 35 years ago, and we have been very happy here."

By then, The Wiremold Company had a network of some 1,700 distributor outlets, agents and direct factory representatives (many of whom had now been affiliated with the company for many years). This network included warehouses in Atlanta, San Francisco and Los Angeles to serve a volume of business that was unimaginable a decade earlier.

When, in the spring of 1965, D. H. Murphy was asked to summarize the spirit of TheWiremoldCompany, he said, "Our management is dedicated to the philosophy that if you have a share in our enterprise, you will want to do your full part in making it successful. The best rule for a businessis the Golden Rule, and together we can make Wiremold a pleasant, profitable place to work in — a place where you can take pride in doing your very best."

Although D. H. Murphy slowed down as he grew

Robert and John Murphy attend an address given by their father to the Newcomen Society at Mystic Seaport in Connecticut on August 5, 1965. An increasingly outspoken opponent of communism in his later years, D. H. Murphy often accepted opportunities like this one, at which he argued for profit sharing as a way to aid capitalism.

older, he never really retired. The Wiremold Company was a grand puzzle that he had been solving for nearly 70 years. He couldn't simply let it go.

Still, by the middle of the 1960s, he had largely withdrawn from day-to-day operations at The Wiremold Company. He remained chairman of the Board of Directors, leaving his creation in the reliable hands of his sons, John and Robert. Looking back on his years at the helm of the company, D. H. reflected, "I have just tried to build a foundation for others to build on."

A senior statesman for business, he turned his energies now to his passion for promoting the free enterprise system by, *New England Electrical News* reported, "getting government out of the businesses it has no constitutional right to be in" and by working closely with a number of organizations committed to advancing capitalism and opposing socialism.

And together, the Murphy brothers presided over The Wiremold Company during one of its greatest periods of sustained growth.

chapter eight

Fostering the Future

"Change will always be with us

if we are to keep pace with

progress. And we intend to do

more than that. We plan to keep

ahead of progress."

— John D. Murphy and Robert H. Murphy in the

program for a company open house, 1968

4000 raceway around the entire perimeter of the building and on interior walls.

By planning the perimeter wiring, nearly 70 percent of the conduit that would have been laid under the building's concrete pad using more traditional and inflexible wiring techniques was completely eliminated. The new building not only provided the administrative space the growing company needed, it exemplified the effectiveness and flexibility of the company's raceways.

The Murphys had already made great strides in the sometimes difficult transition from the simple company their father had nurtured, almost single-handedly, toward the complex organization that it would need to become in order to survive and continue to grow in the future. Still, neither the brothers nor their father could have foreseen how dramatically a new invention would alter the landscape of their industry, nor how swiftly.

As the world moved closer and closer to the new millennium, the pace of change became faster and faster. By the summer of 1969, The Wiremold Company had introduced Tele-Power Poles. In offices coast to coast, the Tele-Power Poles were soon feeding power and telephone wiring from dropped ceiling grids.

Another advance in the integration of power and communication wiring systems for modern offices, the poles would become increasingly popular over the next few years, accommodating a new trend toward open floor plans and the proliferation of more advanced telecommunications and computers. *Electric Ideas* featured interesting applications of this practically "space age" technology.

Toward the end of 1969, *Electric Ideas* published a cover story that began: "In the few years since their introduction, computers have had an enormous impact on our lives. Despite all that has already happened in this dynamic field, however, electrical men should take note of the fact that the best is yet to come.

"As work continues on new generations of computers for the 70s, and as [computer] sales climb toward the expected $16 billion level in 1975 (from $6.5 billion in 1968), there will be increasing needs for efficient electrical distribution systems and increasing opportunities for the men who sell and install them."

Illustrated by a photograph of a ponderous mainframe computer in an office, the article went on to explain how surface wiring had proven to be an excellent way to meet the power demands of computers and pointed to applications of large-sized 3000, 4000 and 6000 series raceways.

The idea that the company's engineering team would, within two decades, become utterly reliant for day-to-day business upon infinitely more powerful "personal" computers that would occupy breadbox space on individual desks would have seemed like, well, science fiction.

But Robert and John Murphy were not afraid to do what had to be done to prepare the company for the technologically challenging future they saw coming at them with frightening speed. And their boldest decision put both their devotion to the company's heritage and their commitment to its future survival in stark relief.

chapter nine

The Right
Ingredients

"We were no longer thought of as just that small company in West Hartford, Connecticut. We were renowned."

— Wayne R. Maschi, retired, 2000

Electrical engineers at the giant Edison Mall in Fort Meyers, Florida, had a challenging assignment in the late 1980s. They chose Wiremold 2100 two-piece raceway with keyless sockets for an elegant lighting system. Some 4,500 feet of raceway were installed to enhance natural lighting from skylights. "Wiremold raceway was the first name that came to mind," said one of the engineers. "It was the high quality product we wanted, and we knew it could accomplish the job."

As the proliferation of desktop computers drove changes in offices, the manufacture of these hi-tech products placed new demands on the manufacturing environment as well. Tele-Power Poles and Plugmold strips figure prominently in this 1975 Hewlett-Packard assembly and testing area.

Tithe impending computer age aside, The Wiremold Company entered the 1970s as dominant in its industry as it had ever been. Every issue of *Electric Ideas* published another story about another innovative application of the company's products by another major corporation. The company could have dropped a name a week. Univac, Upjohn, Montgomery Ward, Bell Laboratories, Westinghouse, Eastman Kodak, and on and on.

Yet in the midst of this growth and the constant development of new product applications, the company published an article in 1971 reminding readers that innovation was not its only strong suit. It had been 55 years since the first raceway, the durable 500 series, had been introduced. Yet that product, and the four that followed it — all introduced before 1936 — were still useful for many jobs. The Wiremold Company wasn't just in the business of making products for the moment. It had a noble tradition of making products to last.

In 1971, the Flexible Duct Divison, renamed in 1970, set up a duct-manufacturing operation in Los Angeles to supplement production in West Hartford. Almost simultaneously, the company opened its own warehouse in Atlanta to more effectively serve the increasing demand for The Wiremold Company's products through the growing Sun Belt. ■

D. H. Murphy continued to come to work daily. Only toward the end, when even his determination and unflagging energy could not lift him, did he relinquish his last hold on the company he had founded.

When he passed away, on March 1, 1973, he was 93 years old. He had guided The Wiremold Company through nearly three-quarters of a century. The incandescent light had been scarcely two decades old when his father bought the Richmondt Electric Wire Conduit Company in Milwaukee. Through some of the darkest moments of the twentieth century, he had never lost hope that the company would become a success. The Wiremold Company was a tribute to his quest for adequate wiring. But perhaps the greatest tribute the company could have paid to him was to roll on without him.

And so it did. By July 1973, the first shipment of the company's new Power Columns had been delivered. A few weeks later, the growing flexible duct manufacturing operation in Los Angeles moved to Pico Rivera, California, where it occupied 19,000 square feet.

In the midst of these changes, the company found itself in need of a new treasurer. A short search was initiated. In May, 1973, just two months after D. H. Murphy's passing, Warren Packard joined the company after 17 years at the Hartford office of the accounting firm Coopers & Lybrand, where he had been a partner.

"I never met D. H. Murphy," he would recall years later. "But I always felt that he was there." ∎

"As a president, John was much like his father," recalls Paul Roedel. "He was a stable leader. He had a level head and he kept the company on an even keel. He was instrumental in ensuring the company remained private when other companies around us were going out of business or being sold."

Indeed, when D. H. Murphy had moved the company to West Hartford, in 1929, the industrial area surrounding it had been home to such companies as AbbottBall, Dunham Bush, Holochrome, Spencer Turbine, Jacobs Manufacturing, Whitlock Manufacturing and Royal Typewriter. By the time John Murphy became president of The Wiremold Company, many of those neighboring companies had either gone out of business or been subsumed by other companies.

John was insistent that this would not be the fate of The Wiremold Company. He felt a strong responsibility to both the Murphy family and the company to advance the vision of his father. And he adhered to that responsibility rigorously.

Increasingly John and Bob began to see the need to recruit leaders from outside the company and build a management team for the future. Though it was still a family company, the days were past when it could be run exclusively by members of the Murphy family. New talent and new ideas would need to be brought in as the company grew and continued to shape itself for a consumer

As the company grew under the leadership of Robert and John Murphy (left and center), it demanded more sophisticated financial leadership, and the Murphy's hired Warren Packard (right) as company treasurer in May 1973, two months after the death of D. Hayes Murphy.

In 1975, The Wiremold Company purchased the Chan-L-Wire overhead lighting system from the Rucker Company of Oakland California. At right, a worker installs Chan-L-Wire, a track lighting system perfect for large retail establishments and factories.

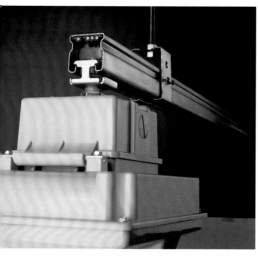

market with increasingly diverse and complex needs.

Up to that point in time, the company's management structure had been organized around two complementary areas: sales/marketing and technical/manufacturing. But now, as the company felt the abrasive impact of an inflationary economy, a third focus emerged. Increasingly, it became clear that the company could benefit from the kind of financial expertise needed to ensure productive customer relations through innovative pricing structures.

In this context, Warren Packard quickly emerged as a leader. By 1974, scarcely a year after he joined the company, he was appointed corporate secretary, retaining the title of treasurer. That same year, as the company's latest innovation — the Overhead Distribution Systems (ODS) was introduced — another plant expansion was opened, adding 46,213 square feet of workspace. ■

On Saturday, October 18, 1975, The Wiremold Company celebrated its 75th anniversary with another open house. In a message to the employees and guests who visited the plant that day, Bob and John said, "As we start on our next 25 years, which will bring us to the century mark, we prepare to meet the challenges that the twenty-first century is already holding out to us."

One of those challenges would be finding increasingly ingenious ways to grow the company. And growth, in turn, would mean diversification by acquiring other companies and other product lines compatible with its own.

Before the 75th anniversary year was over, The

The Wiremold Company experienced unprecedented growth in the late 1970s and 1980s as Wiremold trucks shipped new products across the nation.

Wiremold Company would purchase the Chan-L-Wire system from the Rucker Company of Oakland, California. Like Wiremold raceway, the Chan-L-Wire overhead lighting system was very flexible. Light fixtures could be installed and relocated quickly and easily along the energized strut system. A kind of industrial track lighting, the Chan-L-Wire system was the perfect solution for large retail establishments and industrial factories.

In January 1977, Warren Packard was named executive vice president and a director of The Wiremold Company. Along with this change, John Murphy became chief executive officer and chairman of the board, and Robert Murphy was named president and chief operating officer. Though the company and the national economy continued to be plagued by inflation that dampened both home and auto sales through the 1970s, the company was now sufficiently large and diversified to weather the problem. Net sales that year topped $35,000,000.

But the company had only begun to grow. In 1978, the company accommodated its increasing need for production space by purchasing a plant located in Rocky Hill, Connecticut, from Colt Industries. The flexible duct plant in California, meanwhile — 19,000 square feet when it moved from Los Angeles to Pico Rivera just five years earlier — now occupied 57,000 square feet, and it augured to grow even more.

In many respects, the company had taken on a life of its own. It was bigger than anyone in the Murphy family, including D. H., could ever have envisioned. And it promised to get much bigger still.

So it was that on the cusp of the 1980s, John and Robert Murphy, who had steered the company through this period of remarkable growth and held it on a steady course toward the future, stepped aside to make room for Warren Packard. On May 29, 1979, Packard became the first president and chief executive officer of the company who was not a member of the Murphy family.

John, who had been chief executive officer, remained in the role of chairman of the board. Robert, who stepped down from the presidency, became vice chairman. They continued to be valuable resources to the managers now responsible for the company's future path and retained the fondness and respect of employees. But they turned over management to Packard with full confidence and made it clear to everyone else in the company that they would not second-guess him or interfere with his decisions.

Packard wasted no time in getting to work. He was willing to delegate and anxious to have a management team to which he could quickly turn for counsel as he implemented a slate of new ideas that had been taking form in his mind. He quickly created five new vice presidencies within the company — naming leaders of such key areas as marketing, human resources, sales and engineering — then set to work on a simple and unambiguous agenda: keeping the company private and increasing its profitability.

And under his leadership the company would grow tremendously. Before 1979 ended, the company had purchased the electrical division of Conduits-Amherst Ltd., a Canadian Company, and renamed it Wiremold Canada, Inc., the company's first manufacturing operation outside the U.S. And the flexible duct operations in West Hartford moved to the new Rocky Hill plant.

It was just the beginning of a new period of dramatic growth over which Packard would preside. ■

On the eve of the 1920s, D. H. Murphy's vision had been a business driven by Wiremold raceways. But it was, in fact, loom that kept the business afloat until Wiremold raceway finally caught on, nearly a decade after it first came off the drawing board. By 1979, however, demand for loom had waned significantly. Packard

In spite of all the challenges accompanying a move into a new market, Paul Roedel pushed for development of do-it-yourself products in the early 1980s. DIY proved a strong product area and led to the establishment of the company's plastics operation.

Plugmold Plus plastic raceway, right, is just one line of non-metallic products that grew out of the company's foray into the do-it-yourself market. Plastic products required new manufacturing processes and a leap of faith on the part of the company's management. Easy to cut, as seen opposite, cost-effective and simple to install, plastic raceway was a hugely successful line for The Wiremold Company from its introduction. Plastics now represent the company's fastest growing segment of the business.

was forced to make a difficult decision. He could no longer justify holding on to the loom business, and it was finally sold to Frank D. Saylor & Sons, Inc. of Birmingham, Michigan.

But the early 1980s would bring new lines of business for The Wiremold Company. Thanks to the tenacity of Paul Roedel, the company made its first, tentative forays into the do-it-yourself (DIY) consumer market that, in time, would become a significant component of the over-all market mix.

A born do-it-yourselfer, Roedel had stumbled upon the idea of a DIY product while using the company's products to fix up an old house he had bought during the mid-1970s. " I became convinced there were a lot of guys like me," he recalls. "I was certain there would be a market for do-it-yourself products."

When he tried the idea out on the management team, though, he quickly encountered resistance. DIY, he was told, ran counter to the company's long-standing relationship with distributors. It undercut their market. There was also concern about the potential liability associated with selling wiring products to non-professionals.

Roedel persevered. Over a year, he set up a network of agency representatives through major hardware chains, developed a promotional campaign and got the ball rolling in 1981. The upshot was an increasingly success-ful penetration of the rapidly expanding DIY market.

One important — and entirely unexpected — off-shoot of that new product direction was the develop-ment of raceway products made from substances other than the traditional steel. At European trade shows, more and more competitive products were made of plastic and aluminum. And as competitors began to make inroads in the DIY market with plastic products, Paul Roedel's dogged pursuit of the DIY market became an important factor in the company's decision to begin producing plastic products.

Few changes in the company's history had been so agonized over. "We really resisted getting into plastics initially," recalls Warren Packard. "Historically we had promoted the idea that steel was best. There was great concern that we would damage our reputation with distributors if we started producing plastic products."

For two years, the plastics question occupied a high position on the agenda at annual planning meetings. Company leaders knew what business they were really in — flexible raceway solutions. Could that include plastic raceway? "As far as the customer was concerned, plastic was here to stay," Packard says. The company's long history of developing new products based on customer needs eventually helped it embrace new plastics technology. Trepidation was replaced by relief, as "almost from the moment we began producing plastic products, they were positively received by both con-sumers and distributors," Packard says. In fact, plastics would soon become The Wiremold Company's fastest-growing raceway product line.

The company was in strong ascension. "We just took off," recalls Wayne Maschi. "We were no longer thought of as just that small company in West Hartford, Connecticut. We were renowned."

In 1981, the company's sales volume set an all-time record, topping $50 million. The record was all the more significant because it came at a time when industry sales volume remained essentially flat. Both the housing and automotive markets remained weak, and the sales increase was partially attributable to price increases and new lines of business.

Despite 36 months of decline in the U.S. automotive industry and continued depression of the home construction market, 1982 was another record year for The Wiremold Company. The company continued to relentlessly build for the future, completing Unit 4B, a further extension of the production plant that had been started in 1980.

In June of 1982, Packard announced the company was entering the flat conductor cable market, flat cable line designed for installation of power under carpet tiles. By late summer, the new product line would make The Wiremold Company the only supplier in the world of all three power distribution alternatives — overhead, perimeter and under-carpet. Warren Packard's strategy of diversification was being fulfilled.

Later in 1982, the flexible duct operations were consolidated from Rocky Hill and Pico Rivera to a single operation in Blytheville, Arkansas. And the company grew again, acquiring a controlling interest in Mac Victor Manufacturing, Inc., a Concord, North Carolina, company.

As America grappled with energy shortages, the Mac Victor acquisition seemed to make sense. The company was engaged in development of an energy management system that controlled use of electricity for chain restaurant operations. The energy crisis did not last for long, however, and when it ended, so did the customers' interest in the Mac Victor product.

Shortly thereafter, the company sold its interest in Mac Victor back to the founders. Despite the failure of that particular investment, Packard and the company's management team remained interested in acquisition as a growth strategy.

Only seven years after the company had set a sales record, it wrapped up 1983 just $1 million shy of doubling that record. In fact, 1983 sales represented a whopping 20 percent increase over sales for 1982. It was a milestone year, a year worthy of celebration. But it also marked the end of the run.

By 1984, The Wiremold Company had 2,500 distributors and employed 900 people in seven North American locations. Though the company set another record, the increase over 1983 fell far short of the previous year's dramatic jump. And 1985 sales made an even less dramatic jump, a scant 3.5 percent increase over 1984. Though The Wiremold Company continued to grow, the growth was modest. ■

The overall picture was not a rosy one. The Wiremold Company's increased size and diverse product line, assets in many respects, were turning into liabilities on the factory floor. The company's traditional "batch" manufacturing system relied upon substantial inventories, and as product lines and sales expanded, inventory spiraled out of control.

The Wiremold Company had become incapable of moving quickly. It couldn't produce new products until old products were delivered to customers. Deliveries often lagged. And while the company looked for more space it could acquire for storage, customer service fell off.

The management team knew the company was mired in a sort of status quo, but no one knew exactly what to do to improve the situation. Still, it was clear that unless something was done to shake things up, The Wiremold Company would lose its enviable market position to more flexible, hungry competitors.

In 1985, the management team began to look for answers with the assistance of a consultant, John May, who had worked with the Dexter Corporation on many of that company's successful acquisitions. Packard, Orry Fiume and others met with May for a year, evaluating The Wiremold Company's strengths and weaknesses and various options for pumping new life into the company. In the end, they arrived at two major decisions.

First, they decided to sell the Flexible Duct Division. Once one of the company's strongest areas, it was no longer the dynamo that it had been during the unprecedented growth of the automotive industry following World War II. By the mid-1980s it was consuming a lot of resources and management time, and the auto industry,

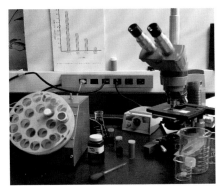

As power quality became a more important issue, The Wiremold Company made two key acquisitions in this area in the late 1980s, Brooks Electronics and Shape Electronics. Sentrex surge protection was one of the new products lines added to the company's offerings.

upon which it was utterly reliant, was less and less predictable. Competitors had emerged, and the division's annual results had become erratic.

Second, with the resources gained from divestiture of the duct business, the company proposed to expand into power and data quality products. More and more electrical appliances were using computer chips, propelling a real concern about the quality of power supplied to them. Surge protection was, the management team recognized, a growth area for the future.

The company moved quickly to sell the Flexible Duct Division in 1986. The sale was followed, in 1988, by the acquisition of Brooks Electronics in Philadelphia, Pennsylvania, and, a year later, Shape Electronics, a Chicago company. Both acquisitions expanded the company's product line into the power quality area. Both evidenced a vision for a company of the future, one that continued to dominate through control of the products the market demanded.

But all too soon, the company would discover that the acquisition of tomorrow's companies was not the panacea it sought. There were deeper problems at the core of The Wiremold Company's lagging growth. And ferreting them out — let alone resolving them — would prove to be a very vexing problem, indeed. ■

One night in 1987, NBC News aired a "White Paper" documentary called "If Japan Can, Why Can't We?" It was a fascinating look at the highly efficient management systems that were making it possible for Japan to beat

America off the mark in head-to-head competition for the rapidly emerging global market. The documentary examined how the Japanese systems worked and asked why the greatest industrial nation in the world could not emulate Japan's new productivity.

One of the millions of viewers who watched the documentary that evening was Orry Fiume. He found the program riveting. He believed, indeed, that he was seeing the future of The Wiremold Company unfold before his eyes.

So impressed was Fiume by what he saw, that he obtained a tape of the program and showed it to the company's executive group. They, in turn, decided to show it to the entire workforce. In shifts, the company's workers went to the old loom space, now vacant, to watch the documentary.

The upshot of this sudden introduction to Japanese manufacturing was, says Fiume, the beginning of the company's understanding that it could no longer rest on its laurels. If The Wiremold Company had prospered until then as the industry leader in its niche, there was no guarantee that it could continue to do so. And if the Japanese model was the wave of the future, then The Wiremold Company clearly had work to do in order to reshape itself for that future.

In fact, the Japanese model resembled closely the Total Quality Management (TQM) philosophy espoused by W. Edwards Deming — a system familiar to Gary Brooks, president of the newly acquired Brooks Electronics. A close acquaintance of Deming, Brooks

had embraced Deming's management model in the early 1980s. He had not only taken his whole management team to Deming's week-long seminars, but had also taken half his entire workforce. Brooks shared his knowledge of Deming's principles with The Wiremold Company's managers.

For a while, it looked as though TQM might be the magic bullet the company needed. On the surface, Deming's management model, summarized in 14 key points, seemed like a good fit for The Wiremold Company. "It fit well with our values, and we all believed in the principles," recalls Fiume. But there was a significant stumbling block. Deming's philosophy was not accompanied with an operational manual. No one really understood how to implement it.

They were determined to try, though. Even Bob Pfund Jr., the company's head of operations, became a believer. An old-fashioned manager, committed to the traditional inventory-based manufacturing model on which the company had always relied, Pfund was not the kind of person one would expect to be an easy convert to Deming. In fact, when the company's flirtation with Deming began, Pfund was negotiating to acquire another building, next door to the West Hartford plant, for more inventory storage space. But after a trip to Japan to explore the management models documented in the NBC report, Pfund came back saying, "We've got too much space."

He could see what the company had to do. It had to reduce inventory and reduce lot sizes. It had to adopt the

The Wiremold Company's board of directors added its first non-family members in the 1980s. Seen opposite are: (top row, left to right) Daniel Murphy, Warren C. Packard, Robert H. Murphy Jr.; (middle row, left to right) Joseph C. Day, John D. Murphy, Richard T. Richardson (son-in-law of John Murphy); (front row, left to right) Millard Pryor, Rosalie Murphy Massey, Marjorie Murphy Morrisey and Robert H. Murphy. This board, with a mix of family members and outside executives, opened the door to manufacturing changes and acquisitions in the late 1980s that would lay the foundation for more dramatic changes in the 1990s.

"just-in-time" (JIT) production model on which the Japanese success was predicated. The trouble was, Pfund didn't know how to do it. And neither did anyone else at the company.

Successfully introducing JIT basically meant reducing inventory. Reducing inventory meant dramatically reducing changeover times — the time it takes to change a machine from setup A for the production of part A to a new setup (setup B) for the production of a different part (part B). Reducing changeover times would mean creating an entirely different sort of production schedule. Pfund admitted he wasn't the man to do it. And he argued for getting someone in operations who could manage the Japanese approach.

Subsequent attempts to emulate the Japanese models — and the nearly disastrous problems engendered by those experiments — preoccupied the company for the next three years. They were, everyone associated with them would agree, mostly a flop. JIT, the company quickly discovered, was more than simply an inventory management system. By the time the company reached this conclusion, however, it was in deep trouble, behind on orders and unable to meet distributors' needs. Some orders went unfulfilled for weeks.

When the company tried to reduce inventory without addressing the company's basic operational philosophy, there was hell to pay. All sorts of manufacturing goblins that had hidden themselves comfortably behind the company's vast inventory suddenly popped out. Its manufacturing system was so antiquated that changeovers

sometimes took days. Without inventory to fill orders while the changeovers were being made, the company was crippled, utterly incapable of moving quickly in any respect.

"It was very bad," says Fiume. "The changes we tried to implement almost sank the company."

To their credit, though, the management did not give up on JIT. To the contrary, these mistakes led to creation of the Wiremold Total Quality Process, outlined in a booklet first distributed to all employees in 1990.

"The Wiremold Company is truly special," Warren Packard wrote in the introduction to the booklet. "For over 90 years, this company has been guided by D. Hayes Murphy's vision which embodies the best of our American system. An uncompromising commitment to customer service has made The Wiremold Company an industry leader. Unshakeable beliefs in job security and in sharing the profits with employees have combined to make the company a preferred employer.

"To survive as an independent and profitable company, we need to focus all of our attention on continuously improving the quality of everything we do — not just the products we make, but also every service we provide, both to our internal and external customers. Until now, you may have had little to say about the quality of materials you received and perhaps less opportunity to be sure what your customer required of you. Now that will change because we have adopted the Total Quality Process — TQP — a process which gives you the means to get exactly what you need from your supplier and the ability to find out precisely what your customer requires from you.

"While there is no question that TQP calls for a big commitment from everyone, The Wiremold Company already has all the right ingredients to make it work: hard working dedicated people, a strong business, and a management team firmly committed to the Total Quality Process...."

The introduction of the Total Quality Process would, in fact, be Warren Packard's final legacy as president of the company. In 1979, he had embarked upon his tenure as the company's leader with a vibrant vision for the kind of company he believed it should become; the kind of company that John and Robert Murphy had charged him with creating.

Now, a decade later, he believed he had made real progress toward that goal. He had led the company through strategic divestitures and acquisitions. And he had played a key role in determining what The Wiremold Company needed to do in order to address its current plateau. He still believed in JIT — wholeheartedly. And just as surely, he realized he wasn't prepared to lead its implementation.

Determined that the company should achieve the vision he and the other members of the management team had been pursuing now for the last half of the 1980s, he took early retirement from active management at the start of 1991 to make room for such a leader. Packard would remain a director of the company until 1999.

The Total Quality Process was part of the company's initial experiment with "just-in-time" manufacturing under Warren Packard's leadership. Lynda Healy, of *The Electrical Distributor (TED)*, presented the NAED Associate Award to Packard at his retirement in 1991 in recognition of his contribution to the industry.

chapter ten

Breaking
Through Walls

"You either change or you don't.

You have to recognize that there

are inherent risks. It's scary. But

so is doing nothing and waiting

for the company to die slowly."

— **Art Byrne, CEO, 2000**

Banish Waste and Create Wealth in Your Corporation, who cited the Wiremold Production System as a success story. "CEOs want to delegate improvement activities, partly because they are timid about going out on the shop floor or to the engineering area or to the order-taking and scheduling departments to work hands-on making improvements. As a result, they never really learn anything about change at the level where it is really created…. If the CEO spends time in real operations learning just how bad things really are and begins to see the vast potential for improvement, he or she will make the right decisions more often."

"He proved himself to be a man of the people," says O'Toole. "His ideas seemed radical, but he took the time to explain them to everyone. He trained every employee in the company, from senior managers to janitors."

Simultaneously, Seyler began working with union officers to introduce them to the kaizen model. As union officers became familiar with the kaizen approach, they gradually realized that it could help the company and that whatever helped the company could, in turn, help employees.

"Probably the thing that impressed everyone the most," says O'Toole, "was what happened on the first kaizen. Art showed up and led the kaizen. He was there on the floor with everyone else on the team. And he stayed there throughout the entire kaizen. When it came time to move things, he rolled up his sleeves and worked with everyone else."

What O'Toole and the other union leaders discov-ered was that Byrne is "a fair man," says O'Toole. "He's willing to listen and he's accessible." Recognizing the potential positive impact of the kaizen approach, the union consented to changes in its agreement with The Wiremold Company so that ordinary work rules could be suspended during kaizens.

As the union warmed up to Art Byrne's management style, so did more and more employees. Meeting the new president and working alongside him during kaizen activities, they quickly discovered he was a man without guile. And the word spread through the company grapevine to those workers who had not yet had an opportunity to meet the man face to face.

The notion of continuous improvement became utterly real when workers saw kaizen teams moving equipment that had been locked in place for years…and then moving it again. And again. Slowly but certainly, as this process produced unarguably positive results, Byrne began to convince the workforce that the system he was introducing would be to their advantage, that he believed in and depended on their contributions to the process and that he was a leader to be trusted.

"We were in the middle of one of the early kaizens and two of our associates came to me and said they thought a particular piece of equipment could work bet-ter if it was moved to a specific location," Byrne recalls. "The equipment was too long, though. It would only fit if they opened a hole through a wall. So I told them to go ahead and do it."

In a company where such a change would have

The kaizen approach replaced the batch manufacturing method with a system of "pull through" production. By shortening the time it takes to ready the machines to produce a new part, one press and one mill are able to produce all the parts for the "supermarket" that supplies the Tele-Power Poles assembly area. The bins are labeled by quantity and part number, and as Material Handler Anna Borges shops the supermarket to supply the assembly area, the press and mill have a simple visual indicator of what parts need to be made.

undergone intensive management scrutiny in the past, Byrne's spur-of-the-moment decision to sanction breaking through the wall was like a breath of fresh air. "It was like a light went on for them," he says. "They realized that they could actually come up with ideas, solutions to problems, better ways to do things, and we could actually implement those ideas."

Perhaps the greatest cumulative impact of Byrne's novel ideas and the breakneck pace at which he introduced them was the sense of invigoration they brought to the company. People were, for instance, astonished when the word got around that he was willing to tolerate mistakes, so long as people learned from them. Gradually, everyone began to relax and learn to enjoy the magic that was taking place.

One day as he walked through the plant, Byrne became engaged in a discussion with two women working on one of the assembly lines. Their rate of production was low, and they told him it was because the product could not be assembled conveniently. It was, they asserted, a flawed design. They challenged Byrne to see for himself if they were right. And he did. He set aside a whole day to work at their assembly station. After eight hours on the job, he agreed that the product needed to be redesigned and selected the problem for a kaizen. More importantly, he affirmed his growing reputation as "a regular guy."

"Art Byrne is the most comfortable, accessible company leader I've ever met," says Orry Fiume. "He's highly intelligent, but he's also very approachable. With

him there is none of the typical egotism that separates business leaders from their employees. I think the workers began to sense that very early.

"He's very active and energetic and he commands respect by virtue of that energy and all that he accomplishes. But he also takes time to listen. He's open to ideas. Most important, he really, genuinely believes that all work is honorable. I think the employees understand that, and they return the respect in kind."

Wayne Maschi, who worked with Byrne for seven years before retiring in 1999, concurs. "He's very down-to-earth. I think he's the kind of guy who could meet with the President of the United States and feel comfortable discussing high level government issues on one day and then could sit down with average guys the next day, have a few beers and feel equally comfortable talking with them about their lives." ■

The growing comfort level with Byrne not withstanding, it would be some time before JIT and the kaizen model became the accepted culture at The Wiremold Company. In fact, two full years would pass, according to Judy Seyler, before the company finally began to settle into a real sense of equilibrium.

"Wiremold had a huge market share, but it was also functioning like it was huge," Byrne recalls. "It was getting beat up because service was declining. It had become lulled into a false confidence based on the working assumption that because it was really good, it didn't have to change. Profitability was low. The strategy was

Art Byrne and Frank Giannatasio instituted a new system to manage inventory. Here Group Leader Mike Abbatemarco places orders from the company's warehouses — physically represented by "kanban" cards — in a central "mailbox." Each slot represents a particular part and unit of time. With a pre-established parts-per-hour rate, group leaders can quickly gauge the demand and workload for each part. Parts are thus "made to order," keeping lead times to a minimum.

annual growth of 2 to 3 percent. Quite simply, the company had become lethargic, and before I could make anything happen here, I had to wake it up."

To do that, he instituted goals that seemed absurd. Instead of 3 percent annual productivity gains, he demanded 20 percent. Instead of lead times that were hovering in the six- to eight-week range, he said he expected to reduce them to one or two *days*.

"We had to move inventory," Byrne says. "We needed the money trapped in inventory to make changes, and we needed to free up the space for the kinds of changes that would allow us to become more productive."

Fully two-thirds of the company's Rocky Hill facility was being used as a warehouse. Two months after coming on board, he announced that he wanted to bring the shipping function back to West Hartford from Rocky Hill. The proposal was met with astonishment by managers, who realized that nearly all of the space at West Hartford not occupied by machinery was occupied with inventory.

The culture of the company was, indeed, "shook up and traumatized" when Frank Giannatasio came on board as manager of operations in the spring of 1992, bringing with him extensive experience implementing the Toyota Production Model — on which the Wiremold Production System is based — at General Motors and Subaru-Isuzu. "The company had gone through the early retirement package. Art had introduced kaizen, but it was still very new. It was my job to get on the floor and make sure the new system worked.

"When we first got involved with changing the company, the problems we tackled were immense," says Giannatasio. "The company had been dependent upon huge inventories for years. If we ran out of any product, it might be out of stock for as much as 60 to 90 days."

One thing the company was not short on was room for improvement. Changing The Wiremold Company required long days, lots of training and many, many kaizens. Work cell by work cell, Byrne and Giannatasio introduced the company to production control, targeting the cumbersome production processes that had nearly

Emilio D'Ascanio picks up an order for the shipping department. The kanban card in D'Ascanio's hand will be slotted in a mailbox for this assembly area, indicating what quantity of finished product was taken off the rack and therefore needs to be replaced.

Kaizen — A Closer Look

For Machine Operator John Williams, who marks his 40th anniversary with the company in August 2000, the kaizen approach triggered memories of a time in the 1970s when he was involved with producing 500 raceway. "We looked at the process and tried to find some ways to not only increase production but make it easier at the same time," he remembers. The experiment involved timing production and analyzing individual production steps. "When they started doing kaizens, I thought it was similar. Anything that improves production makes sense."

"It took the workforce a couple of years to get used to kaizens," recalls Brian Kelly, a group leader in the company's diagnostic tool room. "It was a big change, but we were all impressed by the company's commitment to it. Employees saw that when you're on a kaizen, anything you need is available to you, and they saw that

kaizens work. I think most people here would agree that the changes have been good for the company. Competition is always in the back of your mind. That gives us an edge."

As more and more large-scale problems have been resolved over the past decade, today's kaizens address achieving even higher levels of productivity. No process is ever assumed to be perfect. Everything can always be improved. And the approach to improvement — kaizen — continues to be apllied throughout the organization.

In one recent kaizen, a team was brought together to evaluate the way the cord-ended Plugmold line — an increasingly popular consumer DIY product — was assembled and packaged. Like all kaizens, the team members were assembled from various parts of the company, brought together for the unique perspectives they each brought to the assignment. The team leader was Senior Project Engineer Tony Vargas. The team also included resourceful engineer and co-leader Franklyn Sperry, Product Technician Rachel Chandler, Maintenance Electrician Tony Cascella, CAD design administrator John Chapdelaine and Marketing Communications Manager Don Torrant. Jordan Gibbs, an engineer from a visiting company considering implementing the kaizen system, was invited to join in the process and the team was aided by JIT Facilitator Beverly den Ouden.

While employees are involved in a kaizen, their absence is dealt with in different ways. Sometimes the company simply stops running equipment, sometimes other workers cover the job and sometimes the company schedules around it. But kaizen always takes precedence over everything else.

Like all kaizens, this one began with a team evaluation of the hand assembly situation to determine the objectives. Chandler is the operator of the assembly and packaging process under review, which required the use of twist ties. The packaging process was time consuming, and all bench operators, including Chandler, were at elevated risk of injuries from repetitive movements. The process also required a great deal of time for inspection. The team also sought to make increases in productivity, because — thanks to effective marketing — consumer demand for cord-ended Plugmold products has grown dramatically in recent years.

A consultant working with the team on this kaizen added another goal to challenge the team. He suggested that at no time should there be more component parts on the table than could be assembled in just one hour. The idea was predicated on the "visual factory" concept, which is a central component of kaizen. With many components on the table, it's impossible to tell the stage at which production is progressing. With only an hour's

worth at any location, the progress can be visually gauged.

With their goals in mind, on the first day of the kaizen, the team spent time observing and timing the assembly process and trying out alternative approaches on the floor. They broke the assembly and packaging process down into as many microsteps as possible. They reconvened in a nearby conference room and discussed everything they had observed, brainstorming about possible solutions. Then they went back on the floor, tried some of the suggestions and doggedly measured their effect on productivity.

Eventually the team broke up into sub-teams to examine different steps. One team tackled packaging. Another looked at the possibility of automating the assembly process. Still another tested different tools that might expedite the process.

Over the next four days, says den Ouden, the team experimented with a wide array of possible solutions. They altered the packaging to eliminate the twist ties. They made many different assembly modifications. They mapped out a route for Chandler's material handler that brought him to her table every hour and created a delivery cart with bins containing enough product components for precisely an hour's worth of work. They initiated a labeling system for material handlers that made it easier for material handlers to quickly

and efficiently locate and deliver product components.

"In the end," says Vargas, "we were able to substantially reduce assembly time, raising productivity from 12 to 16 pieces per hour. We made progress, but some of the innovations we explored are still being worked on. I believe we can make this process even smoother. Kaizen never really ends."

Above, Beverly den Ouden times Rachel Chandler at the assembly bench. Opposite, the kaizen team meets nearby to discuss alternative packaging solutions. Clockwise from top are Tony Vargas, Tony Cascella, Rachel Chandler, Beverly den Ouden and Don Torrant.

In addition to changing its manufacturing operations, the kaizen approach was helping The Wiremold Company introduce new products with new customers in mind. Wiremold Access 5000 nonmetallic raceway, opposite, featuring a selection of finishes, including real wood veneers, was designed with great aesthetic appeal yet all the capabilities of multi-channel raceway. Above, this innovative press kit created editorial attention and subsequent sales leads from architects.

sunk the company during the first, tentative experiments with JIT manufacturing during the late 1980s.

The more that employees became accustomed to kaizens, the more the kaizen approach became an accepted part of the culture. "The act of doing it opens things up," says O'Toole. "People begin to see how change can work."

As the employees accepted the challenge Art Byrne had placed before them, they began to produce dramatic results at The Wiremold Company. One of the early kaizens, for instance, aimed to reduce the company's glacial setup times. Over four days, the kaizen team found ways to cut setup for one area from three hours to less than ten minutes.

Over the next two years, practically every piece of equipment in the company was moved repeatedly as new manufacturing configurations were invented and reinvented with kaizen. Gradually, "champions" began to emerge from the workforce, natural leaders who became committed to the success of the new approach and could be recruited to help drive its success.

They are one of the keys to successful introduction of this production model in any company, according to Giannatasio. Another is "a commitment to persist." "At many companies," he explains, "management will try all sorts of solutions. If one doesn't work, they'll discard it and try something else. Nothing ever lasts. With kaizen you don't look for big gains overnight. You look for incremental gains constantly." ■

Byrne's credibility rose as more and more of his "outrageous" predictions became realities.

Within a year, the Rocky Hill warehouse had been closed and all of the shipping operations had been moved back to West Hartford, freeing space for the growth of the company's plastics operation, also located at Rocky Hill, which today is the fastest-growing segment of the Wiremold raceway business.

Throughput time on products plummeted from six weeks to two days or less. The defect rate on products fell by approximately 50 percent each year. And inventories simply melted away as The Wiremold Company learned how to produce things quickly.

In each of the first three years after Byrne joined the company, productivity improved by 20 percent. By 1995, the inventories he had inherited had been reduced by more than 75 percent.

"Today, lead time is hardly ever more than a day on any of our products," says Giannatasio. "Sometimes it's as short as half an hour. And these changes are as much a reflection of rapid change in the marketplace as of what we've been able to accomplish here. Eight years ago, when Art came here, the company was surviving with 90-day lead times. Today, lead times of 10 days or more are quite possibly a loser in what has become an extremely competitive market."

While employees were impressed with the changes Byrne's approach produced, the frosting on the cake was the boost in profit sharing.

"D. H. Murphy was ahead of his time," says Byrne.

"If profit sharing hadn't been here when I came, I would have introduced it. This is the best profit sharing plan I've ever seen, because it's the simplest. Many plans are designed by finance guys and are encumbered with all sorts of conditions, because they don't really want to pay you. I really want us to pay our associates. Profit sharing is a powerful incentive to work hard and make sure the company is successful."

As Byrne's innovations paid off, employees' profit-sharing payouts more than tripled during the first three years after he arrived. The significance of that change could not be missed by anyone in the company. ■

The changes Byrne wrought were not without their downside, though. Before JIT production had become the company's standard operating procedure, shortages occurred as the vast inventory was reduced, straining relations with distributors.

"The Wiremold Company wasn't very nimble," recalls Doug Herberg, director of marketing and materials at Viking Electric in Minneapolis, a veteran distributor who has worked with The Wiremold Company consistently throughout his 24 years in the industry.

"When Art Byrne came to The Wiremold Company, he recognized the company had some archaic operating methods," says Herberg. "Lead time on key items was good, because they had a very big inventory, but on specialty items it was terrible.

"The transition following Art's arrival was painful for us, as it must have been for other distributors. But

ultimately, as Art's changes brought about new thinking and reduced production time, they also addressed distribution problems. Today, they offer a much more coherent package of goods."

By the middle of the 1990s, The Wiremold Company had undergone hundreds of kaizens. And the extraordinary changes were not limited to production. No area of the company's operations were spared the scrutiny of kaizen. Changes in marketing, distribution and new product development were as much responsible for the company's revival as the new production approaches that made it possible for the company to function without inventory.

Thinking lean once again, as it had in the glory days when D. H. Murphy was inventing it product by product, the company had not only learned how to build things quickly and correctly, but it had also reinvented the way in which it decided what to build.

The time to develop new products was reduced from three years to less than six months, as it had adopted a Quality Function Deployment (QFD) model for developing new products. The process means actively listening to customers, devoting the time up front to really understand what their needs are and then distilling that input into the kinds of durable, functional products around which The Wiremold Company's reputation was built in the first place, products that really address the market's needs.

As the company made progress, Byrne brought in more and more new professionals to tackle the rapidly expanding slate of production and marketing initiatives with which he was transforming The Wiremold Company. The expanded team afforded him time to step away from the activities that had occupied most of his energies during his first two years with the company and tackle the task of investing the profits that derived from changes he was making.

The company had doubled in size under Art Byrne's leadership. "The company had lots of problems," Byrne recalls, "but they were not in any way insurmountable problems. We needed growth more than anything. People often look for growth externally. They worry about getting more of the market. They aim to solve the problem through acquisition of other companies. In fact, we've acquired several companies in the past few years. But our first challenge was to fix the core company.

"In my previous experience, every time we fixed manufacturing problems, we created growth. If you have a generally good name in the marketplace, but you're not getting your product out there effectively, then you should start by concentrating on that. The company had been experiencing two to three percent growth annually, simply because everyone believed that was what could be achieved. Manufacturing problems were preventing the company from enjoying a much higher growth rate, as we quickly discovered." ■

With the money earned as the company reshaped its core business and became lean once more, Byrne began focusing on external growth. He developed a growing portfolio of subsidiary companies whose product lines

The Wiremold Company acquired Atlas Cable Tray to complete its range of wiring solutions. The flexibility of its products and its ability to custom design products make it an easy choice for challenging projects. SpecMate cable tray was selected for CNNfn's New York studio, above, and American Airlines Arena, opposite, in Miami. "No one else could do it the way that it needed to be done," said Jim Stephens, the contractor's project manager for the arena project.

The acquisition of Walker Systems in the mid-1990s opened many doors for The Wiremold Company. Walker's products, such as the Walkerdeck Infloor System pictured at right, were a great addition to the company's existing product line, but Walker also included a sales force experienced in building relationships with architects and specifiers. Acquisition of new lines, including Atlas Cable Tray, Walker Systems and the Interlink Cabling System, helped The Wiremold Company effectively reposition itself as the world's preeminent wire and cable management source.

have enhanced existing lines in ways that it would have taken the company much longer to achieve.

No longer simply a producer of top quality wiring products, The Wiremold Company has invented the vanguard of its industry — total wire and cable management. As one company brochure explains, "It's literally a whole new way to wire that enables architects, engineers and contractors to plan early in the design phase how all buildings will be channeled. The company's wire management systems are modular, accessible and expandable, allowing wires and outlets for power, voice and data to be moved or added quickly and easily."

"The acquisitions we've made during the past decade have helped us to completely redefine what The Wiremold Company is," says Vice President of Marketing Ed Miller. "Fifteen years ago, if you'd asked the Murphy family what our business was, they'd have told you we're primarily a raceway business. Today, we're a wire and cable management business. We offer a wide range of solutions.

"Consider the needs of the modern office workspace. It has become a work *station*. In order for it to accommodate the needs of today's business professionals, it must have clean power for a PC, unfiltered power for the remaining electrical requirements, and communications wiring for the telephone, PC modem and local area network. The transition toward that change began with the arrival of personal computers in the early 1980s, followed by local area networks in the late 1980s. By the beginning of the 1990s, everything was being substantially rewired

as PCs found their way to practically every desktop.

"We had no choice but to reinvent ourselves if we were going to remain competitive in our industry. In all of our traditional markets, the demands of emerging technologies were creating new demands that we had to address. We needed in-floor systems, so we acquired Walker Systems, known for their in-floor solutions. We needed overhead systems and saw their growth potential, so we acquired Atlas Cable Tray. We needed poke-thru devices, so Walker in turn acquired RCI, a known thru-floor company. Piece by piece, we have assembled the company we need to be in order to compete in the future. Through acquisitions, we have not only grown the company dramatically, but we've reached the point where we now offer a wire and cable management solution for practically any business application."

Norm Castellani, formerly of RCI, demonstrates the features of a poke-thru device, above. RCI was acquired by Walker to enhance The Wiremold Company's offerings of infloor wiring solutions.

Computers have thoroughly changed society's definition of workspace. Wiremold prewired raceway, offering wiring flexibility for office space, opposite, is just one example of the tens of thousands of products in today's *Wiremold Buyer's Guide*. But customers are not limited to the catalog. Custom-made products, like these curved aluminum raceway installations at Loyolla Marymount Univeristy's business school, are often the result of listening to customers' needs.

Not only did the acquisitions expand the company's product line, they also made it possible for the company to dramatically redeploy its sales force. The acquisition of Walker Systems, for instance, one of the earliest acquisitions during the 1990s, essentially doubled the company's sales force. The Walker sales force had strong relationships with building owners, architects and engineers, groups that were less familiar with The Wiremold Company. With the combined sales force, the company was able to maintain its longstanding relationships with distributors and contractors while establishing relationships with the new customers.

This new focus, in turn, drove the implementation of the company's new product development and QFD processes. "Outstanding, uncompromising customer service is what we're striving for," says Miller. "Our goal is not only to produce a continuous stream of new products that meet the demands of the marketplace, but to do that more quickly than anyone else in the industry. When you can build anything, any time, any day that your customers need it, that's extreme power. Add to that a faster stream of new products. Now, that's hard to beat."

The subsidiary companies that make up The Wiremold Company, located all over the U.S. and in several foreign countries, have made it a powerful global leader and enabled the company to double in size for a second time under Art Byrne's tenure.

Reflecting this remarkable growth, the *Wiremold Buyer's Guide,* descendent of the old catalogues of

products the company used to publish, has grown from a few hundred products just a few years ago, to tens of thousands today. And the *Guide* fairly screams the message about the extraordinary scope of products and services offered by The Wiremold Company in its new incarnation.

"We continue to change to meet your changing needs," reads Art Byrne's message to the reader in a typical edition. "Construction and facilities management grow in complexity every day, and the need for greater flexibility and accessibility in your building's wiring and cabling systems have led to the development of new systems and solutions. Wiring isn't what it used to be and neither are we.

"The Wiremold Company has emerged as the only full source for floor-to-ceiling wire and cable management systems that are more flexible and accessible than traditional wiring methods. In short, we are the *only way to wire*."

"They've created good opportunities in value engineering," says Matt Gold of Midtown Electric in New York City. A distributor who has worked with The Wiremold Company for two decades, Gold recalls the problems that followed Art Byrne's initial efforts to address the company's sluggish manufacturing systems. But he's quick to add that the changes Byrne produced and the company's QFD approach to new product development are working. "They honestly try to evaluate the needs of individual distributors and they've certainly created unique solutions for our customers," he says.

Wiremold raceway is used to distribute power and data cabling to the New York Public Library's Rose Main Reading Room. An extensive renovation completed in 1998 included 4000 steel raceway along the underside of the tables to allow database and Internet access without obstructing the historic design of the room. The project also required custom-made electrical outlets with "a seamless design," according to Julia Doern, project manager for the architecture firm, and Access 5000 in the room's new multi-media center, opposite.

Full Internet Access 21

Full Internet Access 20

Please sign up at the Information Desk in this Room

Upon request you must give this workstation to a disabled person.

"The management is accessible and willing to help us solve problems. In a world of custom products, they've demonstrated the ability to make end users happy." ■

QFD might sound like a logical approach to doing business. After all, every business listens to its customers. It's one thing to listen to customers, however, and quite another to quickly and flexibly produce new products that address their needs. Doing that requires more than a reengineered company that can move quickly and constantly strives for the kinds of improvements that result from hundreds of kaizen activities annually. It also requires a workforce that understands the demands of a QFD-driven workplace.

That's the responsibility of Linda Coveney, the company's QFD Facilitator, who first met Art Byrne when both of them were working for one of the Danaher companies Byrne managed.

The quest for excellence is deeply ingrained in all of the activities Coveney manages. To make sure that every employee in the company understands what QFD is about, a wide range of education and training programs go on constantly.

Several times each year, groups of employees attend a three-day course with in-depth training on team building, problem-solving, the Wiremold Production System, the company's marketing strategy, the overall QFD philosophy and specific methods the company uses to gather the customer input.

In addition, a five-hour class — Wiremold Success

QFD training is an essential element in every associate's career at The Wiremold Company. Using a practice "product," in this case an oversized pencil, Linda Coveney leads QFD training, in which teams examine different aspects of product development, including customer needs.

Training — gives new employees insight into the company's novel approach to doing business and tools to help them become more successful in their jobs quickly. Wiremold Success Training, in fact, was introduced in 1999 in response to an employee survey suggestion.

Even before this, new hires have an opportunity, within days of employment, to view a videotape featuring company managers explaining, in simple terms, what distinguishes the company from its competition.

No one, not even senior managers, is exempt from training in the QFD environment. The company has instituted a program whereby every five years, all personnel go through three existing courses — team-building, problem-solving and production systems — as a "refresher."

The QFD training is really only part of the company's ongoing commitment to employees. "The company has a tremendous commitment to training," says Coveney. "It's really key to our success and the ability to deliver on the promise of what this company has become under Art Byrne's leadership."

To illustrate that commitment, she points to the rotation program used in the Engineering Department. Newly hired engineers spend their first 12 months on the job in manufacturing areas, learning the company, discovering the real implications of the Wiremold Production System and product development driven by QFD. "It's a tremendous commitment," says Coveney, "because the new engineers are not making a contribution to the engineering effort during that time. But in the long run, it's good for the company. By the time they join the engineering team, they understand our culture and they are much better equipped to do their jobs."

Another new initiative, The Wiremold University, aims to help employees become successful, both personally and professionally. In addition to offering courses and programs through the company, it incorporates the program started in the 1960s of providing tuition reimbursement — including the cost of books — for employees taking courses that help them to become more effective members of the team.

From its beginning, The Wiremold Company has always had a strong commitment to its workforce. That commitment is, if anything, more evident today than ever before. All employees are encouraged to be active participants in the ongoing process of making the company better. "Some of our most productive ideas and suggestions

surface through our employee suggestion program and our formal employee feedback process," cites Art Byrne. ■

In 1999, on the eve of the new century, the company won the coveted Shingo Prize for Excellence in Manufacturing. Named for renowned Japanese industrial engineer Dr. Shigeo Shingo, who assisted in the creation of the Toyota Production Systems, the prize honors companies whose manufacturing and business functions and practices stand as best practice models for companies everywhere.

For The Wiremold Company, it reflects all that the company has accomplished in the last decade, changes that, by any measure, are remarkable. But according to Art Byrne, the future of the company is not only in its product lines and its ability to manufacture them quickly and efficiently.

"Businesses are really only collections of people," says Byrne, reflecting on the remarkable changes The Wiremold Company has gone through in the last decade. "One of the things that attracted me to The Wiremold Company in the beginning was the fact that this was, in every respect, a family company. The atmosphere created here by D. H., John, Robert and Warren was extremely positive. This is a really good place to work, with really

One of The Wiremold Company's goals is to be one of the top 10 time-based manufacturers in the world. By earning the Shingo Prize for Excellence in Manufacturing, the company's West Hartford associates have helped the company achieve that goal.

Wire and Cable Management Ideas is the latest in a string of high-impact, idea-driven marketing publications published by The Wiremold Company. This particular vehicle, an outgrowth of *Electric Ideas*, is targeted at architects and specifiers.

good people. People here like each other.

"When you really take a look at the processes we've introduced you see that they all had precedents here. We do a lot to help our associates be successful. D. H. Murphy always cared about his workforce. He was a pioneer in profit sharing and workplace safety. He believed in programs to enhance morale. We use QFD to learn what our customers need, so that we can produce a continuous stream of new products to meet the ever-changing demands of the market. The company introduced *Wiremold Electric Ideas* in 1953 as a way to solicit input from their customers and deliver exciting ideas and solutions. It was a success from the start and it's still a valuable tool for contractors today. Moreover, we've expanded this idea to include a publication geared toward architects and specifiers called *Wire and Cable Management Ideas*.

"In many ways, this company has gone through very significant changes. We've grown. We've retooled the way we do things. We've dramatically enhanced our product line," Byrne adds. "But, in other ways we haven't changed at all. The watchword for The Wiremold Company has always been integrity. D. H. Murphy absolutely and unbendingly insisted upon it. No one here has forgotten that. It's ingrained deeply in the culture of this company. What the company had lost was the sense of urgency that D. H. had when he was building it. Now we've got that back. We've come a long way back toward where D. H. was when he started this company. It's an exciting place to be."

Index